Basic and Clinical Aspects of Neuroscience Vol. 2

Edited by E. Flückiger (Managing Editor),
E. E. Müller and M. O. Thorner

Springer Sandoz
Advanced Texts

Transmitter Molecules in the Brain

Part I: Biochemistry of Transmitter Molecules
Part II: Function and Dysfunction

With Contributions by

G. Fink J. McQueen A. J. Harmar G. W. Arbuthnott
R. Mitchell J. E. Christie

With 45 Figures

Springer-Verlag Berlin Heidelberg New York
London Paris Tokyo

Professor Dr. Edward Flückiger
Pharmazeutische Abteilung
Präklinische Forschung
Sandoz AG
CH-4002 Basel

Professor Dr. Eugenio E. Müller
Dipartimento di Farmacologia
Facolta' di Medicina e Chirurgia
Universita' degli Studi di Milano
Via Vanvitelli, 32
I-20129 Milano

Professor Dr. M. O. Thorner
Dept. of Internal Medicine
School of Medicine
University of Virginia
Charlottesville, Virginia 22908
USA

Volume 1: The Dopaminergic System
© Springer-Verlag Berlin Heidelberg 1985

ISBN-13:978-3-540-13701-6 e-ISBN-13:978-3-642-69950-4
DOI: 10.1007/978-3-642-69950-4

This work is subject to copyright. All rights are reserved, whether the whole or part of the material is concerned, specifically the rights of translation, reprinting, reuse of illustrations, recitation, broadcasting, reproduction on microfilms or in other ways, and storage in data banks. Duplication of this publication or parts thereof is only permitted under the provisions of the German Copyright Law of September 9, 1965, in its version of June 24, 1985, and a copyright fee must always be paid. Violations fall under the prosecution act of the German Copyright Law.

© Springer-Verlag Berlin Heidelberg 1987

The use of registered names, trademarks, etc. in the publication does not imply, even in the absence of a specific statement, that such names are exempt from the relevant protective laws and regulations and therefore free for general use.

Product Liability: The publisher can give no guarantee for information about drug dosage and application thereof contained in the book. In every individual case the respective user must check its accuracy by consulting other pharmaceutical literature.

Preface

This second volume of *Basic and Clinical Aspects of Neuroscience* is devoted to the various transmitter systems of the brain (classical and neuropeptides). In Part I the basic aspects are given, including a critical appraisal of the methods used yesterday and today to describe such neurotransmitter systems. Part II concentrates on the functioning in the body of these transmitter systems under physiologic and pathologic conditions. It goes on to show how neuroendocrine investigations may give insights into the functioning of neurotransmitter systems at least in the hypothalamus, to end with a chapter which assesses very critically the errors and deficiencies of the concepts and techniques used in the attempt to understand the functioning of the brain and the mind.

The editors have been fortunate to have the eight chapters written by a team of investigators working under the direction of Professor G. Fink in the MRC Brain Metabolic Unit at Edinburgh University. We are grateful to him and his colleagues for their work in writing these chapters and for the fine result they achieved. I am grateful for the editorial work done by Professor E. E. Müller (Milan) and Professor M. O. Thorner (Charlottesville), which made this volume possible.

Basle, May 1987

E. Flückiger
Managing Editor

Table of Contents

Part I: Biochemistry of Transmitter Molecules

Introduction: Role of Chemical Neurotransmission in Brain Function
G. FINK

References 4

Classical Transmitters and Neuromodulators
J. K. MCQUEEN

Process of Synaptic Transmission 7
Classification of Synaptic Messengers 7
Dale's Principle 7
Definitions 8
Classical Neurotransmitters 9
Acetylcholine 9
Amino Acids 10
Monoamines 12
Non-classical Transmitters 13
Neuromodulators 14
Characteristics of Neuromodulators 15
Presynaptic Modulation 15
Postsynaptic Modulation 15
Conclusions 16
References 16

Neuropeptides
A. J. HARMAR

Differences Between Peptidergic Neurones and Those Containing
Classical Neurotransmitters 17
One Neuropeptide Gene: Multiple Products 20
Opioid Peptides 23

Distribution and Functions of Neuropeptides 23
Coexistence of Peptides and Classical Neurotransmitters 24
Conclusions 25
References 26

Methods in the Mapping of Neurotransmitter Systems in the Brain
G. ARBUTHNOTT

Techniques Depending on Degeneration 27
Studies Which Illustrate Complete Neurones 29
Methods Which Depend on Axonal Transport 30
Multiple Outputs from One Site 31
Chemical Specification of Neurons 31
Neurotoxins 32
Immunohistochemistry 32
Anterograde Tracing 34
Regional Energy Metabolism 34
References 35

Part II: Function and Dysfunction

Molecular Aspects of Central Neurotransmitter Function
R. Mitchell

Introduction	37
Approaches to the Study of Central Neurotransmitter Action	38
Mechanisms of Neurotransmitter Action	41
References	44

Clinical Relevance
J. E. Christie

The Functional Psychoses	47
Degenerative Disorders and Dementia	51
Conclusions	53
References	53

Normal and Disordered Central Neurotransmitter Function Studied through the Neuroendocrine Window of the Brain
G. Fink

Basic Studies	56
Clinical Studies	65
Summary and Conclusions	71
References	72

Problems and Prospects
G. Fink

Errors and Deficiencies in Concepts	75
Errors in Techniques and Interpretation of Data	75
Prospects	76
References	77

Introduction: Role of Chemical Neurotransmission in Brain Function

George Fink

MRC Brain Metabolism Unit, University Department of Pharmacology, Edinburgh, United Kingdom

The mechanisms of sensation, reason, memory and motor function have preoccupied philosophers, scientists and physicians since the beginning of recorded history [18]. Influenced by Aristotle, Hippocrates and Plato, Galen (130-200 A.D.) proposed that the vital spirits generated in the body were transmitted by the blood from the heart to the brain where they were converted to animal spirits. The waste products of this reaction were wafted through the cerebral ventricles and by way of the infundibulum to the pituitary gland from where the products were expelled into the nose as nasal mucous. The obsession of the ancients with a dominant role of cerebral ventricles in the functions of the brain (Fig. 1), which persisted until the seventeenth century (and even to the present time), is understandable because at autopsy the ventricles were often enlarged and filled with pus or blood and so it was reasoned that they played a dominant role in brain function. Galen's hypothesis was accepted for 1400 years until it was overturned by Richard Lower, an Oxford physician who was first assistant to Thomas Willis, was responsible for most of the dissections of the brain which illustrated Willis' celebrated *Cerebri Anatome,* and subsequently became physician to Charles I. Lower showed by experiment that there was no connection between the pituitary gland and the nose and concluded "whatever fluid is secreted into the ventricles of the brain and goes from there to the infundibulum to the glandular pituitaria distils not upon the palate but is passed again into the blood and mixed with it". Lower's thesis, which was published in 1670 as an appendix to his book *De Origine Cattarrhi* [15], was of course the template for the modern neurohumoral hypothesis of the control of the anterior pituitary gland on which our modern neuroendocrine and psychoneuroendocrine studies are based (see the chapter Normal and Disordered Central Neurotransmitter Function Studied Through the Neuroendocrine Window of the Brain by G. Fink). Although Lower's thesis seems simple, one cannot overestimate the important role played by men such as Lower, Willis, and especially their predecessors, Vesalius and Harvey, in overthrowing the views of Galen which had become entrenched over 1300-1400 years [18]. Our understanding of the brain was greatly advanced by the work of anatomists such as Vesalius, Thomas Willis, Santiago Ramón y Cajal and several clinical neurologists starting with Hughlings Jackson [18]. We know, for example, which areas of the forebrain are concerned with the reception and analysis of visual and auditory stimuli. From the work of Lawrence and Kuypers [14] we know that the primary motor cortex controls the fine movements of the extremities: the movements of the proximal limb and trunk muscles are controlled by nuclei in the

Fig. 1. The medieval cell theory of localisation of the functions of the brain to three chambers: sensation (first chamber), reason, thought and judgement (second chamber), and memory (third chamber). The chambers probably correspond to the lateral, third and fourth ventricles of the brain. After a diagram from the 1506 edition of the **Philosophia pauperium** *by Albertus Magnus (Albert von Bollstädt; 1193-1280)*

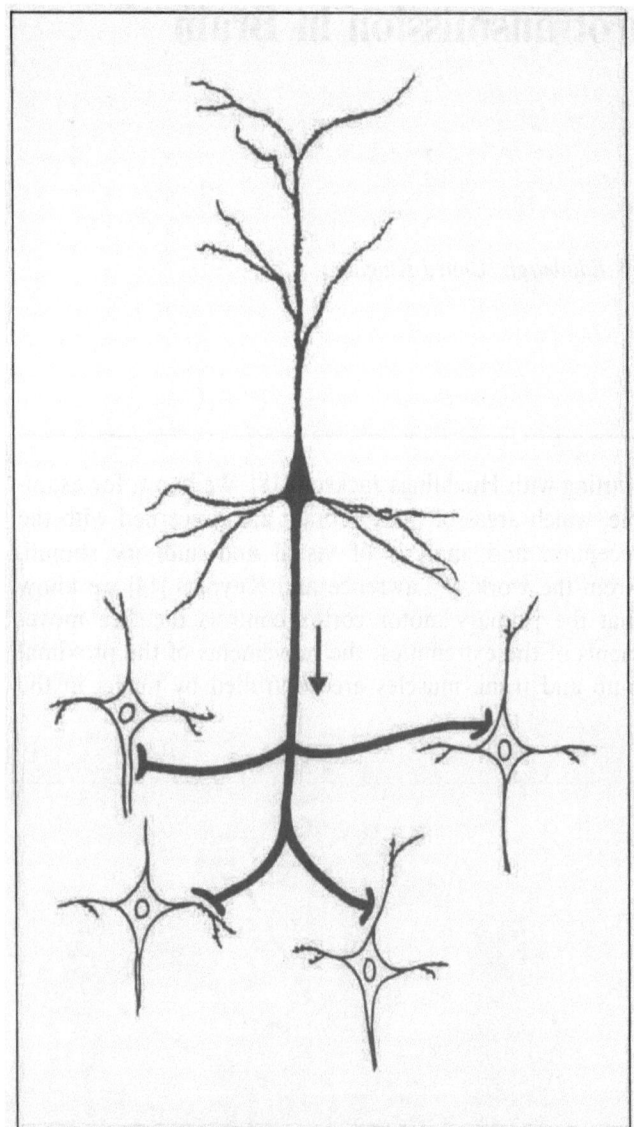

Fig. 2. Schematic diagram of a neurone showing dendrites (with spines) and collaterals of the axon terminating on the dendrites, soma or axon of other neurones. **Arrow** by the side of axon indicates direction of signal

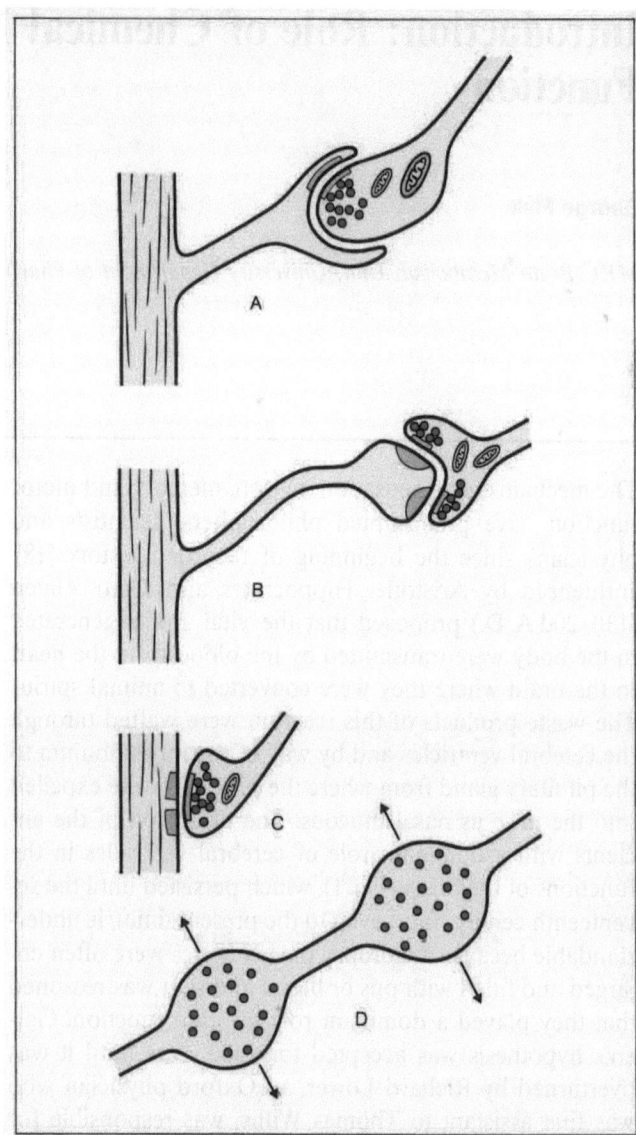

Fig. 3A–D. Schematic diagrams of chemical synapses on dendritic spines (**A** and **B**) and shaft (**C**). The vesicles concentrated at membrane thickenings of the boutons are presumed to contain the chemical transmitter. Many types of central nerve processes, both peptidergic and nonpeptidergic, have varicosities at which, it is believed, release of transmitter can also occur (**D**)

midbrain and hindbrain. However, even though much is known about the sensory and motor system of the brain, our understanding of how the two major systems co-ordinate to produce a coherent response to environmental stimuli seems remote. Perhaps this point is best exemplified by the apparently simple question; what makes us put our foot on the brake of a car when we see a red light? Is there only one or an array of "red-stop" neurones, where is the red-stop neurone and how does it, on receiving a stimulus from the visual cortex, bring about a co-ordinated set of movements that result in the foot being pressed on the brake? However flimsy is our knowledge regarding the motor and sensory system, our understanding of the neurological basis for thought and mood, terms which resist definition, is almost non-existent. Much of our information on the function of the motor and sensory systems of the brain has come from studies of the effects on brain function of lesions due to trauma,

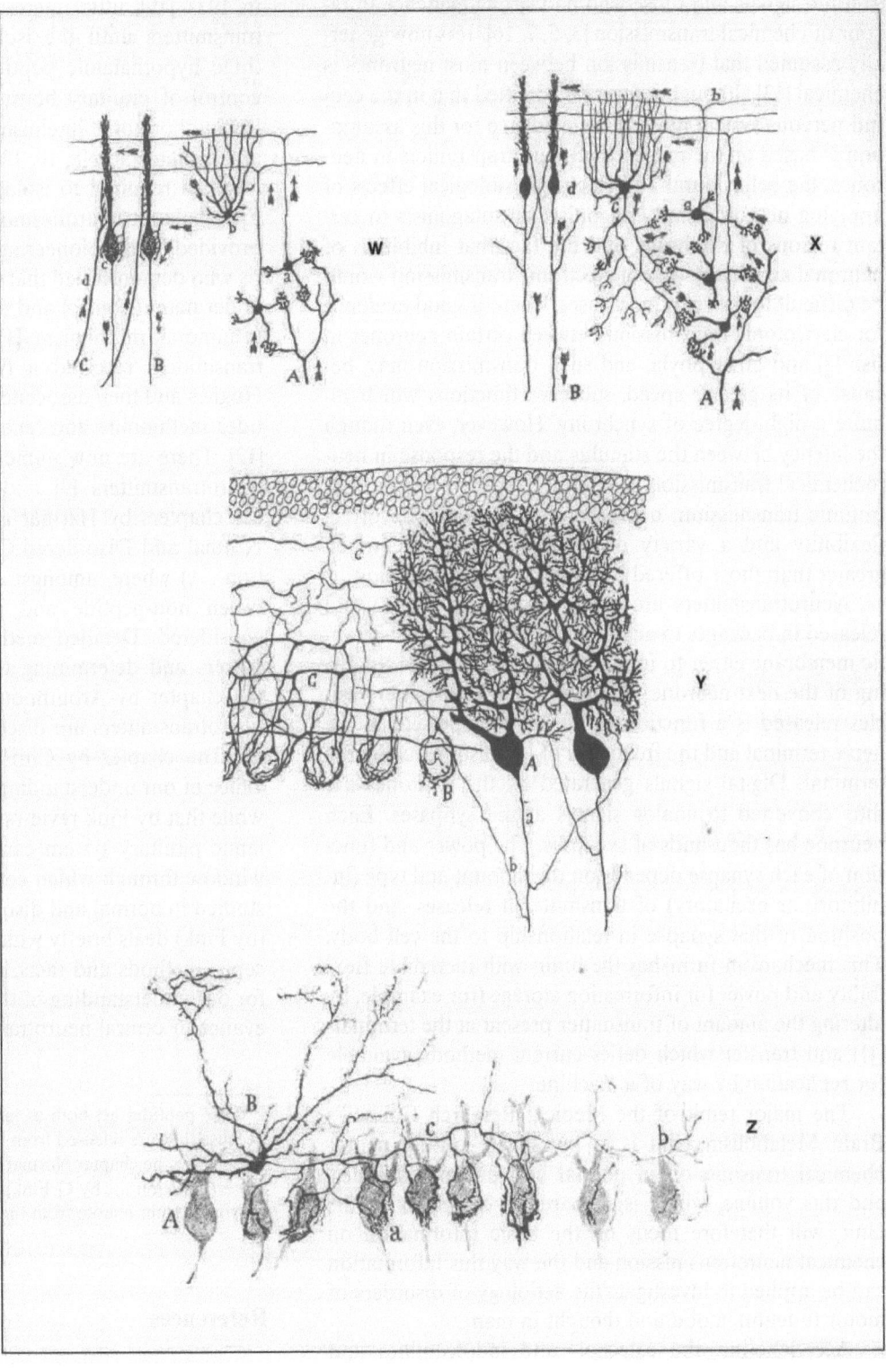

Fig. 4 W-Z. Drawings by Santiago Ramón y Cajal [4] based on Golgi impregnated sections of the cerebellum. The drawings show not only Cajal's technical abilities as a cytologist and draftsman, but also his genius in being able to determine the direction of signals (**arrows in W and X**) from the appearance and connections of the nerve cells and fibres.
W. Shows the way impulses entering the cerebellar cortex by way of mossy fibres (A) affect the firing of granule cells (a), basket cells (b) and Purkinje cells (c). The basket cells inhibit the firing of the Purkinje cells, the output cells of the cerebellar cortex. The firing of Purkinje cells is also influenced by the climbing fibres C. X. Shows that impulses entering the cerebellar cortex by way of the mossy fibres (A) affect the firing of the granule cells (a) which in turn affect the firing of the Purkinje cells (d) and Golgi cells (c). The axons of the Golgi cells terminate on the granule cells. Y. Cross section of a folium of the cerebellum showing Purkinje cells (A), the terminals of basket cells (B) which surround the somata of the Purkinje neurones, and a recurrent collateral (b) of the Purkinje cell axon (a). Z. Shows the way one basket cell (B) in the cerebellar cortex terminates on the cell bodies of many Purkinje cells (A). This small sample of Cajal's drawings illustrates the major contribution Cajal made to our understanding of the complex interactions which can occur between neurones involving the release of both excitatory (e.g. mossy fibre-granule cell synapses) and inhibitory transmitters (e.g. Golgi and basket cell terminals on granule and Purkinje cells, respectively). Cajal's physiological interpretation has been largely confirmed by modern methods [6]

vascular deficiencies or tumours [18], and by analogy we may better understand the neurological basis for normal thought and mood processes by investigating severe psychiatric disorders; that is, the functional and organic psychoses.

Information in the brain is transmitted by way of neurones, specialised cells that have processes termed dendrites and axons (Fig. 2). The flow of information is from dendrite to cell body to axon which terminates on other neurones or, in the periphery, on the neuroeffector cells such as muscle (Fig. 2).

The amplitude of signals passing down the axon is fixed and so changes in the activity of the neurone are reflected in changes in the frequency of impulses. The transfer of information occurs between neurones at specialised junctions termed "synapses". A heated battle raged for many years between scientists who asserted that transmission between neurones occurred by way of electrotonic signals and those who had strong evidence in favour of chemical transmission [3, 5, 7, 16]. It is now generally assumed that transmission between most neurones is chemical [13] although it must be admitted that in the central nervous system most of the evidence for this assumption is based on the presence of neurotransmitters in neurones, the behavioural and electrophysiological effects of applying neurotransmitter agonists or antagonists to certain regions of the brain, and the fact that inhibition of neuronal activity by way of electronic transmission would be difficult to prove in most cases. There is good evidence for electrotonic transmission between certain neurones in fish [2] and other phyla, and such transmission may, because of its greater speed, subserve functions which require a high degree of synchrony. However, even though the latency between the stimulus and the response in neurochemical transmission is much longer than that in electrotonic transmission, neurochemical transmission offers flexibility and a variety of functions which are much greater than those offered by electrotonic transmission.

Neurotransmitters are stored in vesicles (Fig.3) and released in packages to act on receptors on the postsynaptic membrane either to inhibit (Fig.4) or stimulate the firing of the next neurone. The number of transmitter vesicles released is a function of the number present at the nerve terminal and the frequency of impulses reaching the terminal. Digital signals generated by the neurones are thus converted to analog signals at the synpases. Each neurone has thousands of synapses. The power and function of each synapse depends on the amount and type (inhibitory or excitatory) of transmitter it releases, and the position of that synapse in relationship to the cell body. This mechanism furnishes the brain with incredible flexibility and power for information storage (for example, by altering the amount of transmitter present at the terminals [1]) and transfer which defies current methods available for replication by way of a machine.

The major remit of the Medical Research Council's Brain Metabolism Unit is to investigate central neurochemical transmission in normal and disordered states, and this volume, which is prepared by members of the Unit, will therefore focus on the basic information on chemical neurotransmission and the way this information can be applied to investigate the aetiology of disorders of motor function, mood and thought in man.

Acetylcholine, the catechol- and indoleamines and gamma-aminobutyric acid (GABA) are nowadays termed "classical neurotransmitters", first, because, with the exception of substance P, their discovery predates that of the peptide transmitters, and secondly because many of the criteria of a neurotransmitter were based on the electrophysiological action of these transmitters which could be studied relatively easily in the autonomic and peripheral nervous systems [13, 16, 20]. The classical neurotransmitters are the subject of McQueen's chapter entitled Classical Transmitters and Neuromodulators, which is followed by a chapter on the neuropeptide transmitters. Although Von Euler and Gaddum discovered substance P in 1931 [19], little interest was taken in peptide neurotransmitters until the isolation and characterisation of three hypothalamic peptides which mediate the neural control of pituitary hormone secretion; thyrotrophin-releasing hormone, luteinising hormone releasing hormone and somatostatin [8, 10, 17]. The impetus for the intensive research required to isolate and characterise these three hypothalamic neurohormones and neurotransmitters[1] was provided by the pioneering experiments of Geoffrey Harris who demonstrated that the anterior pituitary gland was under neural control and that this control involved a neurohumoral mechanism [11]. Interest in peptide neurotransmitters received a further boost when Kosterlitz, Hughes and their associates isolated the opioid neuropeptides methionine and leucine enkephalin from pig brain [12]. There are now numerous peptides that are putative neurotransmitters [9] and this subject is discussed in the chapters by Harmar and Fink ('Neuropeptides' and Normal and Disordered Central Neurotransmitter Function ...) where, amongst other subjects, interactions between non-peptide and peptide neurotransmitters are considered. Detailed methods for mapping neurotransmitters and determining their function are discussed in the chapter by Arbuthnott, and the modes of action of neurotransmitters are discussed in the chapter by Mitchell. The chapter by Christie discusses the clinical relevance of our understanding of chemical neurotransmitters while that by Fink reviews the way in which the hypothalamic pituitary system can be used as a neuroendocrine window through which central neurotransmission can be studied in normal and disordered states. The final chapter (by Fink) deals briefly with doubts about our present concepts, methods and facts, and considers future prospects for our understanding of the mechanisms and clinical relevance of central neurotransmission.

---

[1] These peptides act both as neurohormones and neurotransmitters because they are released from neurones into the hypophysial portal vessels (see the chapter Normal and Disordered Central Neurotransmitter Function ... by G.Fink) and also at synapses in parts of the nervous system remote from the hypothalamus.

References

1. Bailey CH, Kandel ER (1985) Molecular approaches to the study of short-term and long-term memory. In: Coen CW (ed) Functions of the brain. Clarendon, Oxford, pp 98–129
2. Bennett MVL (1966) Physiology of electrotonic junctions. Ann NY Acad Sci 137: 509–539
3. Blakemore C (1985) The nature of explanation in the study of the brain. In: Coen CW (ed) Functions of the brain. Clarendon, Oxford, pp 181–200
4. Cajal Ramón S (1911) Histologie système nerveux de l'homme et des vertebres, vol 2. Maloine, Paris
5. Eccles JC (1964) The physiology of synapses. Springer, Berlin Göttingen Heidelberg New York
6. Eccles JC, Ito M, Szentágothai J (1967) The cerebellum as a neuronal machine. Springer, Berlin Heidelberg New York

7. Feldberg W (1977) The early history of synaptic and neuromuscular transmission by acetylcholine: reminiscences of an eye witness. In: Hodgkin AL (ed) The pursuit of nature. Cambridge University Press, Cambridge, pp 65–83
8. Fink G (1976) The development of the releasing factor concept. Clin Endocrinol (Oxf) 5: Suppl 245s–260s
9. Fink G (1985) Homeostasis and hormonal regulation (neuroendocrine reflections of the brain). In: Coen CW (ed) Functions of the brain. Clarendon, Oxford, pp 130–159
10. Guillemin R (1978) Biochemical and physiological correlates of hypothalamic peptides. The new endocrinology of the neuron. In: Reichlin S, Baldessarini RJ, Martin JB (eds). The hypothalamus. Raven, New York, pp 155–211
11. Harris GW (1955) Neural control of the pituitary gland. Arnold, London
12. Hughes J, Smith TW, Kosterlitz HW, Fothergill LA, Morgan BA, Morris HR (1975) Identification of two related pentapeptides from the brain with potent opiate agonist activity. Nature 258: 577–579
13. Krnjević K (1974) Chemical nature of synaptic transmission in vertebrates. Physiol Rev 54: 418–540
14. Lawrence DG, Kuypers HGJM (1968) The functional organization of the motor system in the monkey. I The effects of bilateral pyramidal lesions. II The effects of lesions on the descending brain-stem pathways. Brain 91: 1–36
15. Lower R (1670) Dissertation de origine cattarrhi (an addendum to the second edition of De Corde). At the press of J Redmayne, at the expense of Jacobi Allestry, at the sign of the Rose and Crown, St Paul's Churchyard, London
16. Nachmanson D (1959) Chemical and molecular basis of nerve activity. Academic, New York
17. Schally AV, Arimura A, Kastin AJ (1973) Hypothalamic regulatory hormones. Science 179: 341–350
18. Spillane JD (1981) The doctrine of the nerves: chapters in the history of neurology. Oxford University Press, Oxford, pp 1–459
19. Von Euler VS, Gaddum JH (1931) An unidentified depressor substance in certain tissue extracts. J Physiol 72: 74–87
20. Werman R (1966) Criteria for identification of a central nervous system transmitter. Comp Biochem Physiol 18: 745–766

Classical Transmitters and Neuromodulators

Judith K. McQueen

MRC Brain Metabolism Unit, University Department of Pharmacology, Edinburgh, United Kingdom

In the mammalian brain information transfer occurs through the release of chemical messengers or transmitter substances at synapses. For several years it was believed that each neurone contained and released only one chemical transmitter (the so-called Dale's Principle) but it is now known that each branch of a neurone may release several different substances which can influence synaptic transmission. As recently as 1970, only ten substances, acetylcholine, some amino acids and monoamines were thought to be involved in chemical transmission. In 1987 this number exceeds 50, with the discovery in nerve cells of a number of peptides with potential chemical messenger function.

Process of Synaptic Transmission

The pioneering work of Katz and colleagues [18] established the key steps in chemical transmission at the neuromuscular junction. They are fundamental to all synapses and are illustrated schematically in Fig. 1. An action potential travels along the axon and depolarises the presynaptic terminal; ion channels open to allow Ca^{2+} to flow into the terminal and the chemical messenger to be released into the synaptic cleft. At the majority of synapses, interaction of the transmitter with the postsynaptic receptor causes changes in ion permeability. The membrane reaction takes place largely independently of the membrane potential and the channels are called "voltage independent" or "voltage insensitive". If the membrane potential moves towards the threshold for action potential generation, the synaptic action is designated "excitatory". If the membrane potential is stabilised below threshold the action is "inhibitory". At a few synapses ionic gates are not opened but the transmitter promotes a series of enzymatic reactions instead. While the synaptic cleft is being cleared of the transmitter, calcium must also be removed from the cytoplasm, and the most important mechanisms for doing this are: binding to calmodulin; binding to endoplasmic reticulum and other organelles; uptake by mitochondria and efflux by pumping. Levels of free Ca^{2+} are very low in nerve cells (10^{-8}–10^{-6} M) because most is bound, and so small changes may promote significant effects. Calcium plays a crucial role in synaptic transmission and mechanisms controlling the availability of calcium are linked in turn to pathways for transmitter synthesis and cell metabolism. The process of synaptic transmission therefore involves the continual movement of enzymes, ions and the chemical messenger itself between the cell body, the presynaptic terminal, the synaptic cleft and the postsynaptic cell.

Classification of Synaptic Messengers

The uniformity of structure and physiological processes in synaptic transmission is in sharp contrast to the diversity of substances that have been suggested as neurotransmitter substances. These can be divided into three broad categories:
1. Classical or conventional neurotransmitters
2. Putative/postulated transmitters or transmitter candidates
3. Neuromodulators or substances with a modulatory action. Such substances have also been termed "neuroregulators", "neurocommunicators", "neurohormones" or "cotransmitters".

The same chemical might be associated with several different synaptic sites in the brain and might be a neurotransmitter at one, a neuromodulator at another and a transmitter candidate at yet another. These three categories therefore refer to the *action* of the substance rather than the substance itself. Identification of the neurotransmitter at any particular synapse in brain is one of the most difficult problems in neurobiology and consequently the second category is vast. Some authors have chosen to lump transmitter candidates and neuromodulators together so that "modulator" has become one of the most frequently used, and perhaps abused, terms in neuroscience.

Dale's Principle

In 1954 Eccles and co-workers [9] proposed that, based on an interpretation of Dale's writings, "the same chemical

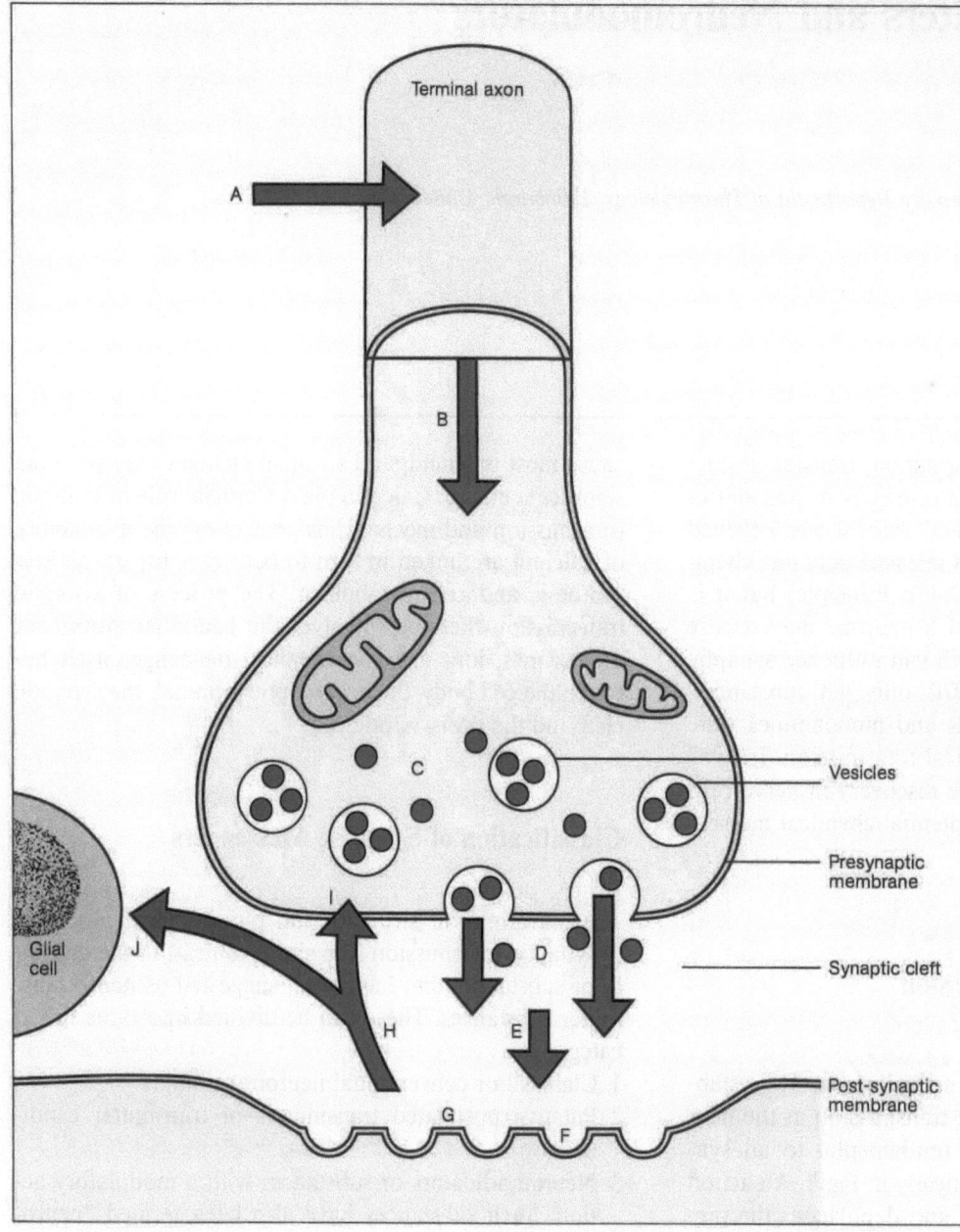

Fig. 1. Process of synaptic transmission. The chemical transmitter or its precursor is synthesised in the presynaptic cell or taken up into the nerve ending by specific transport mechanisms (**A**). Synthesising enzymes (**B**) may convert the precursor into the transmitter, some of which may then be stored in synaptic vesicles (**C**). These enzymes may also be in the synaptic vesicles (e.g. DβH) or in the cytoplasm (e.g. CAT). The action potential which is transmitted along the axon depolarises the presynaptic terminal and ion channels then open to allow calcium ions to flow into the terminal. The key process is Ca^{2+} influx which is believed to promote fusion of the synaptic vesicles with the plasma membrane. The chemical messenger is then released into the synaptic cleft (**D**). Vesicles may act in some cases as stores and transmitter can be released directly across the terminal membrane. The transmitter then diffuses across the synaptic cleft (**E**) and reacts with a specific receptor or binding site (**F**). The sequence of events which follows this interaction may be complex but in the majority of cases there is a rapid and brief opening of specific ion channels in the postsynaptic membrane. While these reactions are occurring, the synaptic cleft is being cleared of transmitter either by enzymic degradation (**G**) or diffusion (**H**), reuptake into the presynaptic terminal (**I**) or uptake by glial cells (**J**)

transmitter is released from all the synaptic terminals of a neurone". Dale [7] had, in fact, been much more cautious and concluded only that the identification of a transmitter at the peripheral endings of a sensory neurone might "furnish a hint as to the nature of the transmission process at a central synapse". Unfortunately, the use of the term "principle" and the authority of Dale's name elevated a working hypothesis to a law which had two important implications. First, each neurone could contain only one transmitter substance; secondly, a neurone would secrete the same transmitter at all of its endings. The single transmitter idea was consistent with available evidence at the time. However, studies in vivo and on nerve cells in culture have shown that this is now untrue. Several substances with potential for chemical messenger function can coexist in the same neurone. Therefore single-transmitter status is not a principle of neuronal cell biology. In addition, there is no conclusive evidence that all branches of a neurone secrete the same transmitter. Dale's Principle is, therefore, not a law but rather – as Dale himself saw it – of "at least some value as a stimulus to further experiment".

Definitions

A classical neurotransmitter can be defined as a chemical substance which is synthesised in a neurone, contained in the presynaptic terminal, released in response to nerve activity, and acts at specific specialised receptor sites on the postsynaptic cell to produce changes, usually in ionic conductances, which alter the activity of the recipient cell.

Certain criteria have been proposed to distinguish transmitters from other physiologically significant substances, and procedures have been Table 1 near here devised for their identification [11, 20], as set out in Table 1.

Substances for which some evidence is still lacking are termed "putative transmitters" or, if the information is speculative, "transmitter candidates". The most difficult criteria to satisfy are those of *"release"* and *"identity of action"* due to difficulties in access to synapses in the CNS and to limited availability of selective pharmacological agents. The recent vast increase in the variety of chemical substances thought to be neurotransmitters in brain is largely due to the development of new methodologies, such as immunohistochemistry, recombinant DNA technology, cell culture.

When Florey [12] introduced the concept of "modulator substances" as a distinct group acting together with "neurotransmitters", he perceived the nervous system as being composed of different neurochemical types of neurones, each manufacturing and storing only one transmitter substance. Any other chemicals found within nerve cells, could not therefore be transmitters. We now know that neurones contain several substances which may have transmitter actions. Florey [12] used the term "modulator substance" for "any compound of cellular and non-synaptic origin that affects the excitability of nerve cells" and that "can affect the responsiveness of nerve cells to trans-synaptic actions of presynaptic neurones". This idea has now been extended so that neuromodulators are seen as important compounds in general communication between nerve cells and which operate more like hormones. They need not necessarily be non-synaptic. Unlike neurotransmitters, they are not restricted spatially to the subsynaptic membrane or indeed temporally to the duration of the presynaptic action potential. More importantly, on their own, neuromodulators have no effects on postsynaptc excitability but they can alter the responsiveness to the classical neurotransmitter.

Neuromodulators may also influence the release, binding or other action of a neurotransmitter. Modulatory roles have been ascribed to different substances on the basis of such diverse characteristics as: the (complex) nature of the electrophysiological response, the (slow) time course of action or the mode of transmission to target cells.

Classical Neurotransmitters

As a general rule, neurotransmitters are small molecules synthesised by specialised enzyme systems. They are found in moderate concentrations in brain (nanograms to micrograms per gram) and show high affinity for specific receptor sites. These factors have been utilised in the development and application of new methodologies which have provided data to confirm a range of compounds as classical neurotransmitters in mammalian brain.

The first substance to be established as a neurotransmitter was acetylcholine (ACh) and 5%-10% of brain synapses are thought to be cholinergic. Amino acids (γ-aminobutyric acid -GABA, glycine, glutamate, aspartate and taurine) form the main group of central transmitters (estimated 25%-40% synapses; see Snyder [23]). The monoamines (catecholamines, tryptamines and histamine) account for only 1%-2% of brain synapses but the monoamine pathways, which have been extensively characterised using fluorescence histochemistry, are widely distributed throughout the brain.

Acetylcholine

The blood-brain barrier allows only glucose, essential amino acids, ions and fatty acids to enter the brain. Therefore, most neurotransmitters must be synthesised from precursors taken up from the blood stream (see Fig. 2). Dietary choline is taken up into cholinergic neurones by a specific transport system and ACh is produced

Table 1. Criteria for establishing a substance as a neurotransmitter

Criterion	Detail	Methodology used
Presence	1. The chemical must be localised in presynaptic elements, and probably unevenly distributed in the brain 2. Precursors, synthetic enzymes or a specific transport system must be present in the neurone 3. Specific binding sites or receptors for the chemical should be present at the synapse	Biochemical Anatomical Histochemical
Release	Stimulation of afferents should cause release of the chemical in physiological amounts	Physiological
Identity of Action	1. Direct application of the chemical to the synapse should produce effects identical to those produced by physiological stimulation 2. Interaction of the chemical with its receptor should induce membrane changes leading to excitatory or inhibitory postsynaptic potentials 3. The effects of afferent stimulation or of direct application of the substance should be similarly affected by pharmacological agents	Physiological Pharmacological
Removal	Inactivating mechanisms (e.g. diffusion, metabolising enzymes, reuptake system) should exist to terminate interactions of the substance with its receptor	Biochemical Histochemical

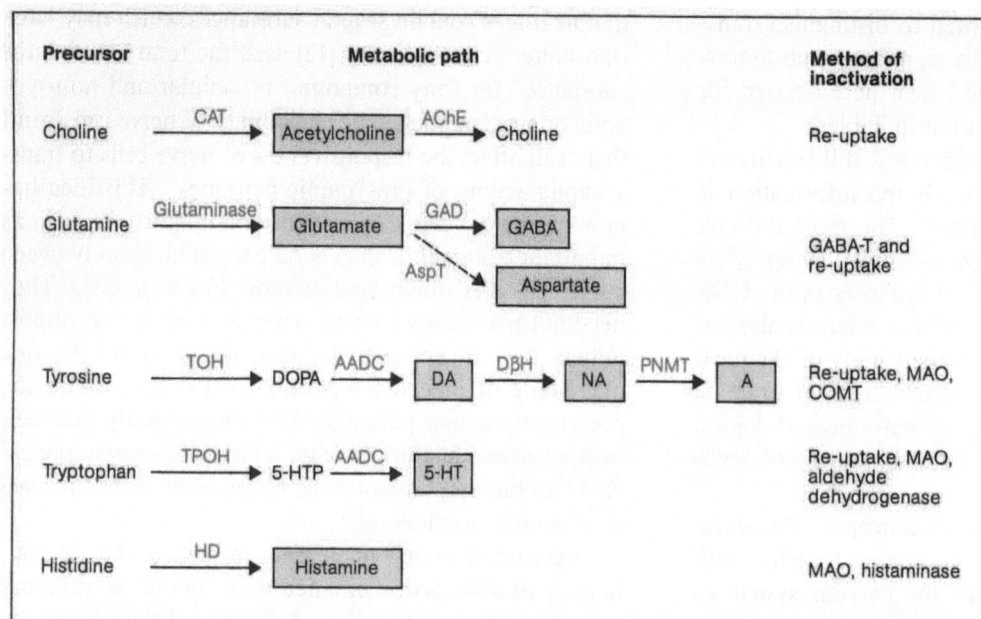

Fig. 2. Steps in the formation of classical neurotransmitters. *A*, adrenaline; *AADC*, amino acid decarboxylase; *AChE*, acetylcholinesterase; *AspT*, mitochondrial aspartate transaminase; *CAT*, choline acetyltransferase; *COMT*, catechol-O-methyltransferase; *DβH*, dopamine β-hydroxylase; *DA*, dopamine; *DOPA*, dihydroxyphenylalanine; *GABA-T*, GABA transaminase; *GAD*, glutamic acid decarboxylase; *HD*, histidine decarboxylase; *5-HT*, 5-hydroxytryptamine; *5-HTP*, 5-hydroxytryptophan; *MAO*, monoamine oxidase; *NA*, noradrenaline; *PNMT*, phenylethanolamine-N-methyltransferase; *TOH*, tyrosine hydroxylase; *TPOH*, tryptophan hydroxylase

by the enzymic action of choline acetyltransferase (CAT) and stored in vesicles. Following synaptic release, ACh is broken down by the enzyme acetylcholinesterase (AChE) into choline, which is taken up again into the presynaptic terminal to replenish the stores. AChE histochemistry was used 25 years ago to map central cholinergic neurones. Subsequently, specific antibodies to CAT were used to trace cholinergic pathways more precisely. More recently, an antibody to ACh itself has been described [10] and its use may facilitate the complete localisation of cholinergic systems. Large cholinergic cell groups are found in several sites in mammalian brain: in the medial septal nucleus, the diagonal band, the nucleus basalis magnocellularis and the pontine nuclei. Axons project from these cell bodies to the hippocampus and throughout the cerebral cortex (Fig. 3). ACh released from the axon terminals acts as a ligand and binds to specific sites in a receptor protein to produce a change in membrane permeability. Two types of muscarinic site and also nicotinic sites (based on the pharmacological characteristics of receptors in the peripheral nervous system) have been described in several regions throughout the brain. The type of response is determined by the receptor on the postsynaptic cell, but ACh is predominantly an excitatory neurotransmitter.

Amino Acids

The most widely utilised inhibitory neurotransmitter in the mammalian CNS appears to be *GABA* [8] with *GABA*-accumulating nerve endings constituting up to 50% of total nerve endings [17]. The excitatory amino acid glutamate is converted into inhibitory GABA by the enzyme glutamic acid decarboxylase (GAD) (Fig. 2). The actions of synaptically released GABA are curtailed by GABA-transaminase. Because GABA is involved in neuronal intermediary metabolism (the "GABA shunt") levels of GABA do not accurately reflect the density of GABAergic innervation. GAD has been purified, and specific antibodies to it have been produced. The resultant immunohistochemical studies using these antibodies have provided maps of GABA-containing neurones (Fig. 3). GABA is present in a number of classical inhibitory systems such as small interneurones in the dorsal horn of the spinal cord and the Purkinje, basket and stellate cells of the cerebellum. However, GABA is also a neurotransmitter in longer projection systems such as those between corpus striatum and substantia nigra, and between sub stantia nigra and ventromedial nucleus of the thalamus. GABA's actions are fast (milliseconds), and it interacts with two types of binding site (designated "high" or "low" affinity). These receptors control the ionophore or conductance channel for chloride ions which move during a GABA inhibitory postsynaptic potential (IPSP). Some drugs, e.g. picrotoxin and bicuculline, act at specific sites on the protein complex to block control of the Cl^- ionophore by GABA. Benzodiazepines act at different sites to produce facilitation of the GABAergic synapse. GABA is cleared from the synaptic cleft by uptake into the presynaptic terminal and into glia, which recycle GABA through glutamate and glutamine back into the nerve ending. *Glycine* is also an inhibitory neurotransmitter in interneurones in the spinal cord, where its detailed physiological role has been elucidated. There is little evidence that glycine affects neurones above the brainstem.

Glutamic acid, or glutamate, and *aspartate* are excitatory amino acids which are also constituents of intermediary metabolism. There is considerable physiological evidence that glutamate and/or aspartate mediate brief excitatory actions at several sites in brain, e.g. striatum, hippocampus and, particularly, granule cells in the cerebellum (Fig. 3).

Fig. 3. Principal locations of neuronal pathways. *AM*, amygdala; *ARC*, arcuate nucleus; *DCN*, deep cerebellar nuclei; *DH*, dorsal horn; *DRG*, dorsal root ganglion; *EPN*, entopeduncular nucleus; *GP*, globus pallidus, *HAB*, habenula; *HIP*, hippocampus; *HYP*, hypothalamus; *LC*, locus coeruleus; *LTA*, lateral tegmental area; *MED*, medulla; *MSG*, medullary serotonin group; *NA*, nucleus accumbens; *OB*, olfactory bulb; *OT*, olfactory tubercle; *PBA*, parabrachial area; *PC*, pyriform cortex; *PERI-V*, periventricular grey; *SC*, superior colliculus; *SCN*, suprachiasmatic nucleus; *SEP*, septum; *SN*, substania nigra; *STR*, striatum; *THAL*, thalamus; *VP*, ventral pallidum; *VTA*, ventral tegmental area. Modified from Shepherd [22]

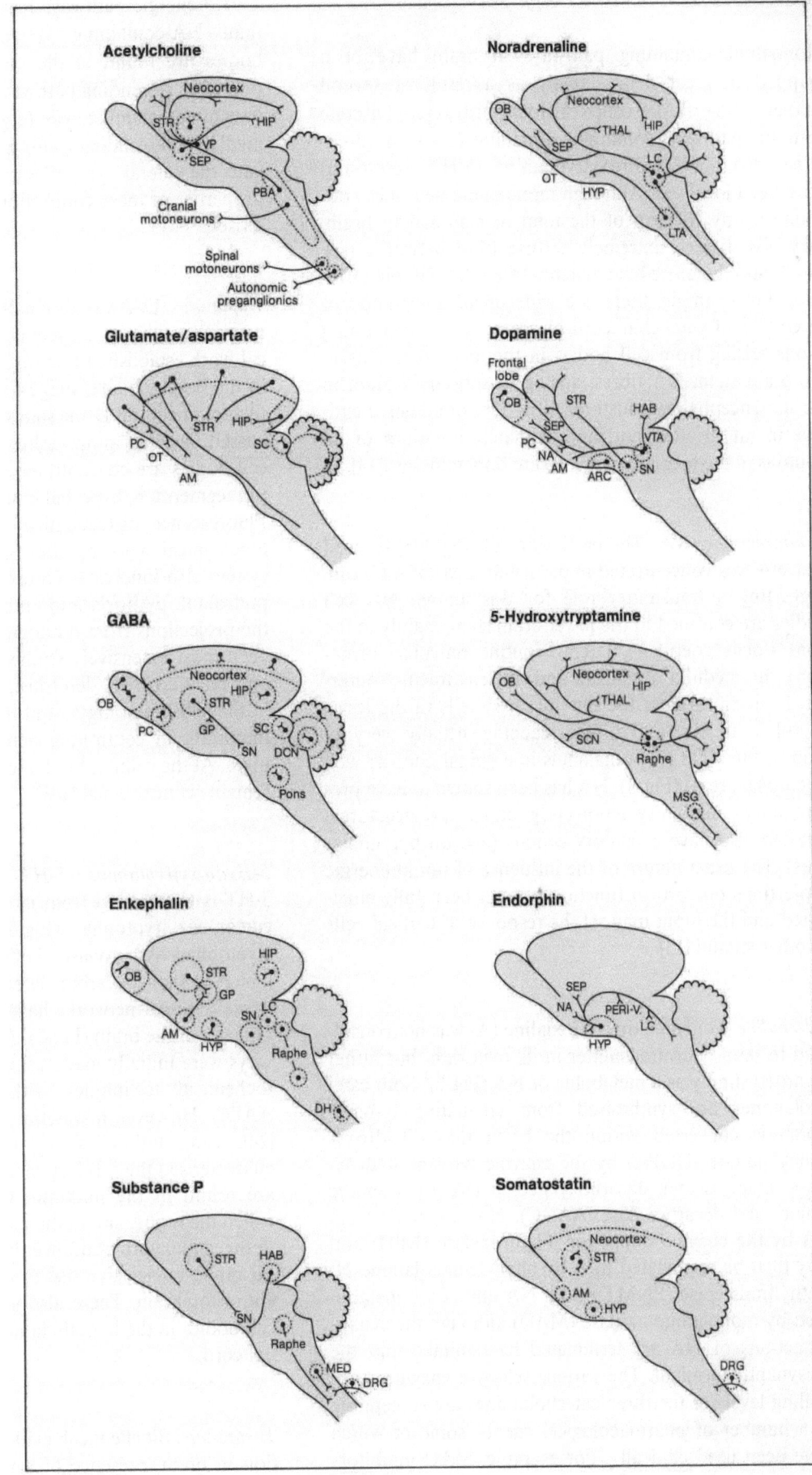

Monoamines

Monoamine-containing pathways in brain have been mapped in great detail largely as a result of several studies using fluorescence histochemistry [2]. Discrete neuronal pathways containing dopamine (DA), noradrenaline (NA) or 5-hydroxytryptamine (5-HT, serotonin) have been localised. Although monoamine neurones constitute a tiny fraction of the total in mammalian brain, they give rise to extremely diffuse fibre networks (see Fig. 3) and therefore have potential for considerable influence. For example, there is a widespread innervation of cerebral and cerebellar cortical areas by NA-containing axons arising from cell bodies in the locus coeruleus in the brain stem. DA fibres, arising in substantia nigra in the mesencephalon, constitute 10%-15% of all nerve endings in rat striatum and approximately one-third of all boutons in the median eminence are dopaminergic [14].

Noradrenaline (NA). The early work of Vogt [26] showed that NA was concentrated in particular areas of rat brain, suggesting a transmitter role for this amine. NA cell bodies are confined to the lower brain stem, mainly in the pons (locus coeruleus, lateral pontine reticular formation), the medulla oblongata and nucleus tractus solitarius. Projections from the few hundred cells in the locus coeruleus are long and diffuse, reaching virtually every region in the CNS (hypothalamus and spinal cord as well as cortical areas) (Fig. 3). NA has been shown to have predominantly inhibitory actions (e.g. on cortical structures) but can also have excitatory actions (e.g. on hypothalamus). The exact nature of the influence of noradrenergic projections on cortical function has not been fully elucidated and the input may set the response of cortical cells to other stimuli [13].

Adrenaline. For some time adrenaline (A) was not considered to be a neurotransmitter in its own right but rather regarded simply as a metabolite of NA (Fig. 2). Both catecholamines are synthesised from circulating tyrosine which is converted within the brain into dihydroxyphenylalanine (DOPA) by the enzyme tyrosine hydroxylase. Dopa is then decarboxylated to DA (by aromatic amino acid decarboxylase, AADC). NA is formed from DA by the enzyme dopamine β-hydroxylase (DβH) and may then be methylated to A by phenylethanolamine-N-methyltransferase (PNMT). Both NA and A are metabolised by monoamine oxidase (MAO) although the synaptic actions of NA are terminated by reuptake into the presynaptic terminal. The various selective enzymes controlling levels of the three catecholamines are susceptible to a number of pharmacological agents, some of which have been used clinically. For example, MAO inhibitors have been used as antidepressants, and peripherally acting decarboxylase inhibitors used in conjunction with L-DOPA to treat Parkinson's disease.

Adrenergic pathways in the brain are more restricted than NA-containing systems. PNMT-containing cell bodies are found in the medulla oblongata and lower pons [15]. Ascending pathways reach central grey and various hypothalamic nuclei (e.g. paraventricular and dorsomedial). Descending pathways to the spinal cord innervate the lateral sympathetic column. The physiological properties of these connections have not yet been characterised.

Dopamine. DA was originally thought to be present in neurones solely as a precursor of NA (Fig. 2). Biochemical work established that high concentrations of DA exist in the basal ganglia, suggestive of a transmitter role. The picture of central DA systems today is complex and is discussed in detail in an earlier volume [14]. DA-containing cell bodies are concentrated in the substantia nigra, ventral tegmentum, hypothalamus and olfactory bulb (Fig. 3). Fluorescence histochemistry has confirmed the earlier biochemical work of Glowinski's group [25] that a DA system also innervates cortex, in particular the cingulate, prefrontal, pyriform and entorhinal areas. However, it is the projections from the nigral DA neurones which have been most extensively studied. DA appears to be stored and released from dendrites, rather than axon terminals in the substantia nigra, and this dendritic release may be important in controlling neuronal activity in substantia nigra. At the cellular level, the nature of the actions of DA remains controversial [14].

5-Hydroxytryptamine (5-HT). Like the catecholamines, 5-HT is synthesised from its circulating amino acid precursor, viz tryptophan (Fig. 2). This is hydroxylated by tryptophan hydroxylase (TPOH) to 5-hydroxytryptophan which is then decarboxylated to 5-HT. Extensive 5-HT nerve terminal networks have been described in virtually all parts of the brain (Fig. 3). Attempts to trace 5-HT pathways were initially made using autoradiographic and histochemical techniques with antisera to TPOH and AADC. However, it was the work of Steinbusch's group [24], using antibodies to 5-HT itself, which eventually characterised the 5-HT systems. In the rat, 5-HT neurones are found in the midbrain located in clusters of cells called the raphé, and in the medulla oblongata (projecting to the spinal cord and lower brain stem). Cells in the dorsal raphé nuclei give rise to extensive projections to the entire forebrain. There also seem to be 5-HT-containing cell bodies in the hypothalamus, substantia nigra and spinal cord.

Histamine. Biochemical evidence (e.g. uneven distribution in brain, presence of synthesising enzyme histidine decarboxylase) suggests that histamine may be a central neurotransmitter. However, it remains in the putative category because a confirmatory histochemical technique which would enable histamine-containing neurones to be localised is not yet available. Recently, Watanabe et al. [27] have used antibodies to histidine decarboxylase to map a restricted hypothalamic systems with cell bodies in the caudal magnocellular hypothalamic nuclei.

The time course of action of monoamines is slower than that of ACh or amino acids and may involve a series of enzymatic reactions rather than direct opening of ionic channels. In one such mechanism, the transmitter is bound to a membrane protein linked to an adenylate cyclase. This activates an internal receptor, a protein kinase which promotes protein phosphorylation and eventually results in a change in the ionic conductance of the membrane. So-called second messenger systems, first described for the hormonal action of insulin, are discussed in detail in a later chapter in this volume (Mitchell). In contrast, steroid hormones act by a different mechanism. They affect a cytoplasmic or nuclear receptor protein and the hormone receptor complex produces its specific effects on nuclear DNA. This is important not just during development to control the expression of male and female characteristics, but also in adulthood.

Adenosine and Related Compounds. Although there is some evidence that adenosine can be released together with manoamines and can stimulate cyclic adenosine monophosphate (cAMP) formation, perhaps producing long-lasting changes in cellular activity, these compounds have not been convincingly established as central neurotransmitters.

Non-classical Transmitters

During the last 15 years a large number of small peptides has been localised in central neurones. Snyder [23] anticipated in 1980 that their number might eventually exceed 200 and this is unlikely to be a gross overestimate. In contrast to the classical neurotransmitters, peptides are small- and medium-sized molecules which are present in very low concentrations. Their synthesis is slow and complex; they are formed from large polypeptide prehormones in the rough endoplasmic reticulum via a smaller prohormone in the Golgi body. The neuropeptides overlap in size between classical transmitters on one hand and hormones on the other. Indeed several in the hypothalamus can be considered as neurohormones, since they are released either into the hypophyseal portal vessels (e.g. luteinising hormone releasing hormone [LHRH], thyrotropin-releasing hormone [TRH], and somatostatin [SS]) or the systemic circulation (vasopressin and oxytocin). Neuroendocrine cells are the first to appear in the evolution of the primitive nervous system and similar peptides appear across a range of species from hydra to man. The key differences between peptidergic and other neurones are discussed in detail in chapter 3.

Whether or not neuropeptides are true chemical transmitters is the subject of continuing intense research. In general, the criteria of presence and removal have been satisfied and there is increasing evidence for release. However, establishing "identity of action" is proving much more difficult.

Many of the peptides which are hypophysiotropic hormones are not restricted to the neuroendocrine system, i.e. hypothalamus and pituitary. For example, LHRH, TRH and SS have been found in the cerebral cortex, midbrain and spinal cord, as well as the peripheral nervous system. The brain also contains pituitary hormones such as adrenocorticotropic hormone (ACTH), oxytocin and vasopressin; and gut hormones, such as cholecystokinin (CCK) and vasoactive intestinal peptide (VIP). There is an oxytocinergic and vasopressinergic innervation of the nucleus tractus solitarius, the habenula and septum, and both peptides are released from terminals at these target areas in a calcium-dependent (i.e. transmitter) fashion.

Besides neuroendocrine peptides, a large number of other peptides identified in brain in recent years must now be considered as transmitter candidates. The enkephalins, other opioids and substance P have been intensively studied and there is now convincing evidence favouring their inclusion in the category of neurotransmitters [6]. The two opioid peptides (methionine enkephalin and leucine enkephalin) are present in discrete systems in the CNS [16] which are different from those containing β-endorphin [3]. A third opioid system, containing dynorphin-like peptides has also been described.

Immunohistochemical techniques have been used to map the distribution of various peptide neurones (Fig. 3). Some brain areas are rich in both cell bodies and terminals. The cerebellum and thalamus have low levels of most peptides but cortical areas are particularly rich in peptides; e.g. VIP, CCK and SS have been found in predominantly non-pyramidal cells.

The multitude of peptides found in the CNS raises the question of their relationship to neurones containing classical transmitters. There are now many examples of coexistence of peptides and classical transmitters in the same neurone, and this may represent the rule for peptides rather than the exception. The functional significance of this type of coexistence in CNS neurones and the potential for simultaneous release from the same nerve endings is not yet clear. Possible mechanisms of release and interaction are indicated in Fig. 4. In the periphery, peptides seem to support the action of the classical transmitter in some sites, e.g. VIP and ACh in salivary gland. However, not all branches of the same classical system may contain the same peptide: for example, CCK is present in DA neurones only in the mesencephalon; other DA neurones contain neurotensin. Peptides may not convey the rapid transmission of impulses but may rather participate in the regulation of long-term events by exerting trophic actions or influencing other intracellular events. Alternatively, peptides may only be released when neurones are firing at

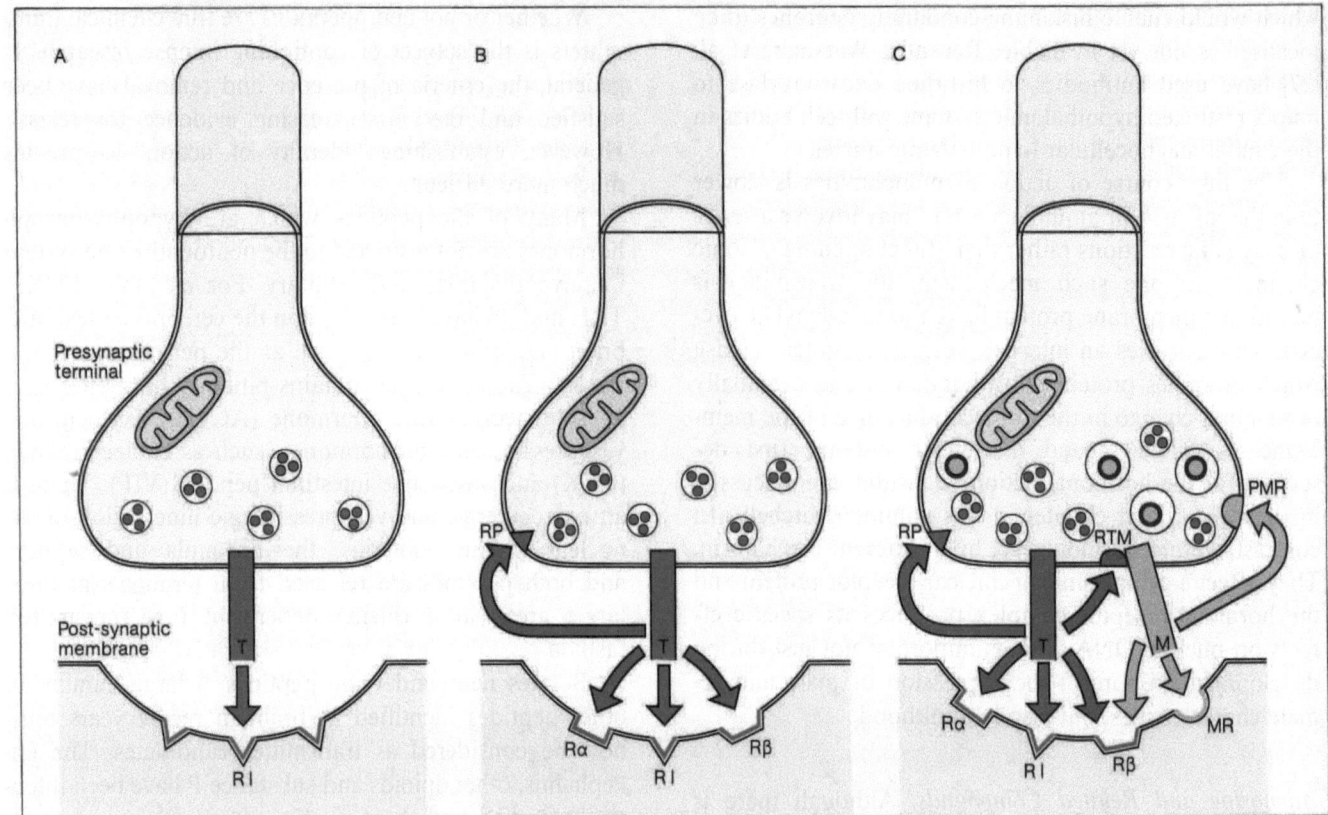

Fig. 4A-C. *Some mechanisms of release and interaction of multiple transmitters (small vesicles containing classical transmitter T; larger, dense-core vesicles containing transmitter T and a neuropeptide M). In A, T acts on a single postsynaptic receptor RI. In B, T acts on multiple types of postsynaptic receptor RI, Rα, Rβ and on a presynaptic autoreceptor RP to control its own release. In C, T and M are both released. T can inhibit the release of M at a presynaptic receptor RTM; M acts on its own postsynaptic receptor MR and at a presynaptic receptor PMR to modulate synaptic transmission. Modified from Lundberg and Hökfelt [19]*

a high rate. Release of small amounts of more than one chemical transmitter may also avoid up or down regulation of receptors.

The one peptide for which there is particularly convincing evidence that it is a neurotransmitter is substance P, the first peptide isolated from the brain. The major substance P circuits in the CNS of experimental animals have been mapped using tract-tracing procedures and immunohistochemistry (Fig. 3). Substance P is found in several short projection tracts, e.g. striatonigral, habenulo-interpeduncular and in spinal cord. It is a more potent excitant than glutamate on dorsal root ganglion cells where its physiological action has been studied in detail. Although its time course of action may be slow this need not necessarily exclude it from a role as a neurotransmitter.

In general, the effects of neuropeptides on cell potentials are slow and this has led many to assume they can only have a modulatory function in controlling the responses to rapidly acting classical neurotransmitters. However, it is possible that in another 15 years we shall not only know of the location of many more peptides, but we shall also know more of their precise mechanisms of action.

Neuromodulators

"Modulation" is defined in the dictionary as "the alteration of a primary characteristic over time by a secondary influence". Whereas the term "synaptic transmission" describes both the event and the mechanism involved, the term "synaptic modulation" does not make a precise statement. Indeed, modulation is used in several different contexts to describe actions at different levels of organisation: from the synapse to the behavioural level to refer to changes in response produced by motivation, arousal or learning.

The roots of the concept of modulation are in endocrinology and the actions of hormones: Florey [12] introduced the term "modulator substances" in 1967 to describe compounds of "cellular and nonsynaptic origin that affect the excitability of nerve cells" and which can "affect the responsiveness of nerve cells to transsynaptic actions of presynaptic neurones". Although peptides, such as angiotensin, ACTH and prolactin have their main physiological actions in the periphery, they can have behavioural actions like neurohormones. Steroid hormones act on specific target sites within the CNS and cause long-latency, long-duration alterations in neuronal function. This genomic action of steroids may underlie many of their effects in the CNS. Some workers [1] see neuromodulators as compounds important in general communication between nerve cells, operating rather like hormones. However, the term has also been used loosely to refer to putative transmitters, e.g. taurine. In particular, modulation is being used increasingly to characterise actions that do not fit easily into the traditional concepts of neuronal action. For example, neuropeptides have often been classified as neuromodulators or neuroregulators partly because of their long-lasting, complex effects.

Characteristics of Neuromodulators

1. They have no effects on their own but can alter the effects of neurotransmitters (i.e. contingent action).
2. Longer and slower time course of action.
3. Neuromodulators are not restricted spatially to the synapse or temporally to the duration of the presynaptic action potential. Whereas neurotransmitter release is dependent on orthodromic nerve impulses occurring in the neurone which produces the substance, modulatory substances may be released in a continous or intermittent secretion process, e.g. from glia.
4. Modulators may have intracellular actions (rather than effects on membrane receptors).

Presynaptic Modulation

The amount of neurotransmitter released into, and retained in, the synaptic cleft can be altered by a neuromodulator acting at various stages in the transmission process. No fewer than six possible sites of intervention are indicated in Fig. 1 (B, C, D, H, I and J). Both presynaptic facilitation and inhibition have been described and are discussed in detail by Chesselet [5]. Many neurotransmitters can alter their own release through auto-receptors (Fig. 4B) and this effect can be described as modulatory. Cyclic AMP has been shown to facilitate Ca^{2+} influx and may therefore regulate transmitter release from neurones in the CNS. Conversely, it is thought that opiates suppress the effects of substance P on sensory nerve terminals by decreasing Ca^{2+} influx. Reducing Ca^{2+} influx has been suggested as the mechanism underlying presynaptic inhibition in several systems.

Adenine compounds, cAMP, adenosine diphosphate (ADP), adenosine triphosphate (ATP) and adenosine have a depressant effect on synaptic transmission [21]. Based on the results from intracellular recordings and reports that adenine compounds reduce the amount of neurotransmitter released, the site of action is presumed to be presynaptic and a high-affinity Al adenosine receptor has been characterised on presynaptic terminals. Adenine compounds are released from the dendrites and axon terminals of central neurones. In the hippocampus it has been shown that adenine compounds affect both excitatory and inhibitory circuits thus altering the balance between the two.

Postsynaptic Modulation

Modulators can alter the affinity or number of receptors (Fig. 1 F) or change the time course and extent of the permeability changes induced by transmitter action. For example, opiates alter the coupling of the glutamate receptor to its ionophore and benzodiazepines produce allosteric alterations in the GABA binding site.

In general, neuropeptides modulate the actions of classical neurotransmitters with which they coexist, through postsynaptic modulation. For example, Braitman and his colleagues [4] have studied the effects of TRH on ACh and glutamate responses in the CNS. TRH potentiates the slow excitatory response to ACh, but only on pyramidal tract neurones. In contrast, TRH depresses the responses to glutamate on non-pyramidal tract cells. Although the mechanism is unknown, the receptors involved probably exist on different sets of neurones.

Another example of the complexity of postsynaptic modulation concerns substance P, which inhibits the spontaneous activity of most spinal neurones but excites some. Substance P selectively depresses nicotinic ACh responses but has no effect on muscarinic or glutamate responses. Enkephalins, which can excite or inhibit cells directly may, in the absence of a direct action, potentiate or depress the effect of other transmitters. These are only a few examples of the many interactions, classed as modulatory, which have been described between peptides and neurotransmitters.

Changes in synaptic function induced by the modulatory action of chemical substances can occur by a variety of physiological and biochemical processes. Such effects might be the means of tuning or co-ordinating the activities of vast arrays of target neurones. The most common features of the neuromodulator are long duration of action and contingent action. These properties are ideally suited to regulating the state of behavioural responsiveness of an organisms, e.g. through the control of learning, arousal, etc.

Conclusions

The development of sensitive histochemical and biochemical methods has led to the discovery in the brain of large numbers of chemical substances which appear to have a transmitter function. Of particular interest has been the finding that a single neurone may produce and release more than one chemical messenger at its synapses. Some substances fit into the category of classical or conventional neurotransmitters, fulfilling criteria originally devised for ACh's action at the neuromuscular junction. Others are termed "putative transmitters" or "transmitter candidates"; future advances in methodology may allow some members of this group to be reclassified. Some of the many neuropeptides found in the CNS appear to be true neurotransmitters but are in many ways different from the smaller classical neurotransmitters; other neuropeptides may act as neuromodulators, substances which have no effect on their own but can dramatically alter the response to the actual transmitter. With the continued use of the techniques of molecular biology in neuroscience, new understanding of the mechanisms of neuronal communication becomes possible. Given the potential for chemical switching at synapses, the complex mixtures of pre- and postsynaptic receptors, and tissue-specific synthesis of two or more peptides, it is apparent that the flow of information between neurones is a continually dynamic process. More than ever before, the synapse must be considered as a functional unit for information processing rather than just a structural entity.

References

1. Barchas JD, Akil H, Elliott GR, Holman RB, Watson SJ (1978) Behavioural neurochemistry: neuroregulators and behavioural states. Science 200: 964-973
2. Björklund A, Hökfelt T (1984) Handbook of chemical neuroanatomy. Vol 2 Classical transmitters in the CNS, Part 1. Elsevier, Amsterdam
3. Bloom FE, Battenberg E, Rossier J, Ling N, Guillemin R (1978) Neurons containing β endorphin in rat brain exist separately from those containing enkephalin: immunocytochemical studies. Proc Nat Acad Sci USA 75: 1591-1595
4. Braitman DJ, Auker CR, Carpenter DO (1980) Thyrotropin-releasing hormone has multiple actions in cortex. Brain Res 194: 244-248
5. Chesselet M-F (1984) Presynaptic regulation of neurotransmitter release in the brain. Neuroscience 12: 347-375
6. Cuello AC, Priestley JV, Sofroniew MV (1983) Immunocytochemistry and neurobiology. QJ Exp Physiol 68: 545-578
7. Dale HH (1935) Pharmacology and nerve-endings. Proc R Soc Med 28: 319-332
8. De Feudis FV, Mandel P (1981) (eds) Advances in biochemical psychopharmacology. Vol 29 Amino acid neurotransmitters. Raven, New York
9. Eccles JC, Fatt P, Koketsu K (1954) Cholinergic and inhibitory synapses in a pathway from motor-axon collaterals to motoneurons. J Physiol (Lond) 126: 524-562
10. Eckenstein F (1985) Antibodies to acetylcholine at last. Nature 318: 236
11. Elliott GR, Barchas JD (1979) Neuroregulators: neurotransmitters and neuromodulators. Behav Brain Sci 2: 423-424
12. Florey E (1967) Neurotransmitters and modulators in the animal kingdom. Fed Proc 26: 1164-1178
13. Foote S, Bloom FE, Aston-Jones G (1983) Nucleus locus coeruleus: new evidence of anatomical and physiological specificity. Physiol Rev 63: 844-914
14. Fuxe K, Agnati LF, Kalia M, Goldstein M, Andersson K, Harfstrand A (1985) Dopaminergic systems in the brain and pituitary. In: Fluckiger E, Muller EE, Thorner HO (eds) Basic and clinical aspects of neuroscience. Springer, Berlin Heidelberg New York Tokyo, pp 11-25
15. Hokfelt T, Fuxe K, Goldstein M, Johansson O (1973) Evidence for adrenaline neurons in the rat brain. Acta Physiol Scand 89: 286-288
16. Hughes J, Smith TW, Kosterlitz HW, Fothergill LA, Morgan BA, Morris HR (1975) Identification of two related pentapeptides from the brain with potent opiate agonist activity. Nature 258: 577-579
17. Iversen LL, Bloom FE (1972) Studies of the uptake of [^{3}H] GABA and [^{3}H] glycine in slices and homogenates of rat brain and spinal cord by electron microscopic autoradiography. Brain Res 41: 131-143
18. Katz B (1966) Nerve, muscle and synapse. McGraw-Hill, New York
19. Lundberg JM and Hökfelt T (1985) Co-existence of peptides and classical neurotransmitters. In: Bousfield D, (ed) Neurotransmitters in action. Elsevier, Amsterdam.
20. McLennan H (1963) Synaptic transmission. Saunders, Philadelphia
21. Schubert P, Lee K, Kreutzberg GW (1982) Neuronal release of adenosine derivatives and modulation of signal processing in the CNS. In: Buijs RM, Pevet P, Swaab DF (eds) Progress in brain research, vol 55. Elsevier, Amsterdam, pp 225-238
22. Shepherd GM (1983) Neurobiology. Oxford University Press, Oxford
23. Snyder SH (1980) Brain peptides as neurotransmitters. Science 209: 976-983
24. Steinbusch HWM (1981) Distribution of serotonin-immunoreactivity in the central nervous system of the rat-cell bodies and terminals. Neuroscience 6: 557-618
25. Thierry AM, Stinus L, Blanc G, Glowinski J (1973) Some evidence for the existence of dopaminergic neurons in the rat cortex. Brain Res 50: 230-234
26. Vogt M (1954) Concentration of sympathin in different parts of central nervous system under normal conditions and after administration of drugs. J Physiol (Lond) 123: 451-481
27. Watanabe T, Taguchi Y, Shiosaka S, Tanaka J, Kubota H, Terano Y, Tohyama M, Wada H (1984) Distribution of the histaminergic neuron system in the central nervous system of rats; a fluorescent immunohistochemical analysis with histidine decarboxylase as a marker. Brain Res 295: 13-25

Neuropeptides

Anthony J. Harmar

MRC Brain Metabolism Unit, University Department of Pharmacology, Edinburgh, United Kingdom

Since the early 1970s, it has become clear that in addition to the "classical" neurotransmitters, many small peptides are present in nervous tissue (Fig. 1) and neurotransmitter or neuromodulatory roles have been proposed for several of them [2, 7].

The first neuropeptides to be fully characterised were the nonapeptides vasopressin and oxytocin, which were isolated from the posterior pituitary gland by Du Vigneaud [15] and his colleagues in the early 1950s. Soon afterwards, the search for factors released from the hypothalamus, which control the release of anterior pituitary hormones, led to the identification of peptides capable of stimulating the release of thyrotrophin (thyrotrophin-releasing hormone), luteinising and follicle-stimulating hormones (luteinising hormone releasing hormone - LHRH) and of inhibiting the release of growth hormone (somatostatin). Within the last 5 years, hormones stimulating the release of adrenocorticotrophic hormone corticotrophin-releasing factor) and growth hormone (growth hormone releasing hormone) have been characterised. Attempts to purify these releasing factors have led to the fortuitous discovery of other neuropeptides including substance P and neurotensin.

Many neuropeptides were first discovered as hormones in peripheral tissues and only shown subsequently, by immunohistochemical and radioimunoassay techniques, to be present in nervous tissue. Examples of this class of neuropeptides include vasoactive intestinal polypeptide (VIP) and cholecystokinin. A further group of neuropeptides has been predicted from molecular biological studies of the polypeptide precursors to other known peptide hormones. Calcitonin gene-related peptide (CGRP), a product of the mammalian gene encoding the hormone calcitonin, and PHM-27, a peptide found within the amino acid sequence of the human VIP precursor, are examples of this type.

Differences Between Peptidergic Neurones and Those Containing Classical Neurotransmitters

The mechanisms by which neuropeptides and classical neurotransmitters are synthesised and utilised in neurones are strikingly different [4]. Classical neurotransmitters such as noradrenaline, are synthesised predominantly in nerve terminals, as the result of enzymatic transformations of the small molecules which are their precursors (Fig. 2). The enzymes which catalyse these transformations (for example dopamine β-hydroxylase - DβH - a key enzyme in the biosynthesis of noradrenaline) are, like all cellular proteins, synthesised in the neuronal cell body by the ribosomal translation of messenger RNA (mRNA). Many of these enzymes are sequestered within vesicles which, after axonal transport to nerve terminals, become neurotransmitter-containing secretory granules. The mRNA for DβH, like those for all proteins destined for secretion from the cell, encodes a precursor polypeptide containing at its N-terminus a hydrophobic "signal" sequence of amino acids which is responsible for the translocation of the newly synthesised polypeptide chain through the membrane of the rough endoplasmic reticulum into structures which ultimately become secretory vesicles. The signal sequence is cleaved from the growing polypeptide chain. As newly synthesised secretory vesicles pass down the axon and accumulate in nerve terminals, they acquire their content of neurotransmitter through the activity of DβH and of the other enzymes of noradrenaline biosynthesis. After release from nerve terminals, noradrenaline may be inactivated by oxidative enzymes or may be recycled by reuptake into nerve terminals. Thus, the stores of noradrenaline in nerve terminals are maintained by reuptake, by synthesis of transmitter within secretory granules and by axonal transport of newly synthesised granules into nerve terminals.

The pathways by which the neuropeptides are synthesised closely resemble those for DβH and other cellular proteins [3]. Translation of peptide mRNA in the neuronal cell body leads to the synthesis of a precursor polypeptide containing at its N-terminus a signal sequence which is cleaved from the precursor during its synthesis (Fig. 2). As the precursor passes from the endoplasmic reticulum into the Golgi apparatus and is packaged into secretory vesicles, further proteolytic cleavages take place, typically at sites flanked by pairs of basic amino acids (arginine and lysine), liberating the mature peptide from its precursor [14]. These proteolytic events, together with enzymatic modifications of individual amino acids within

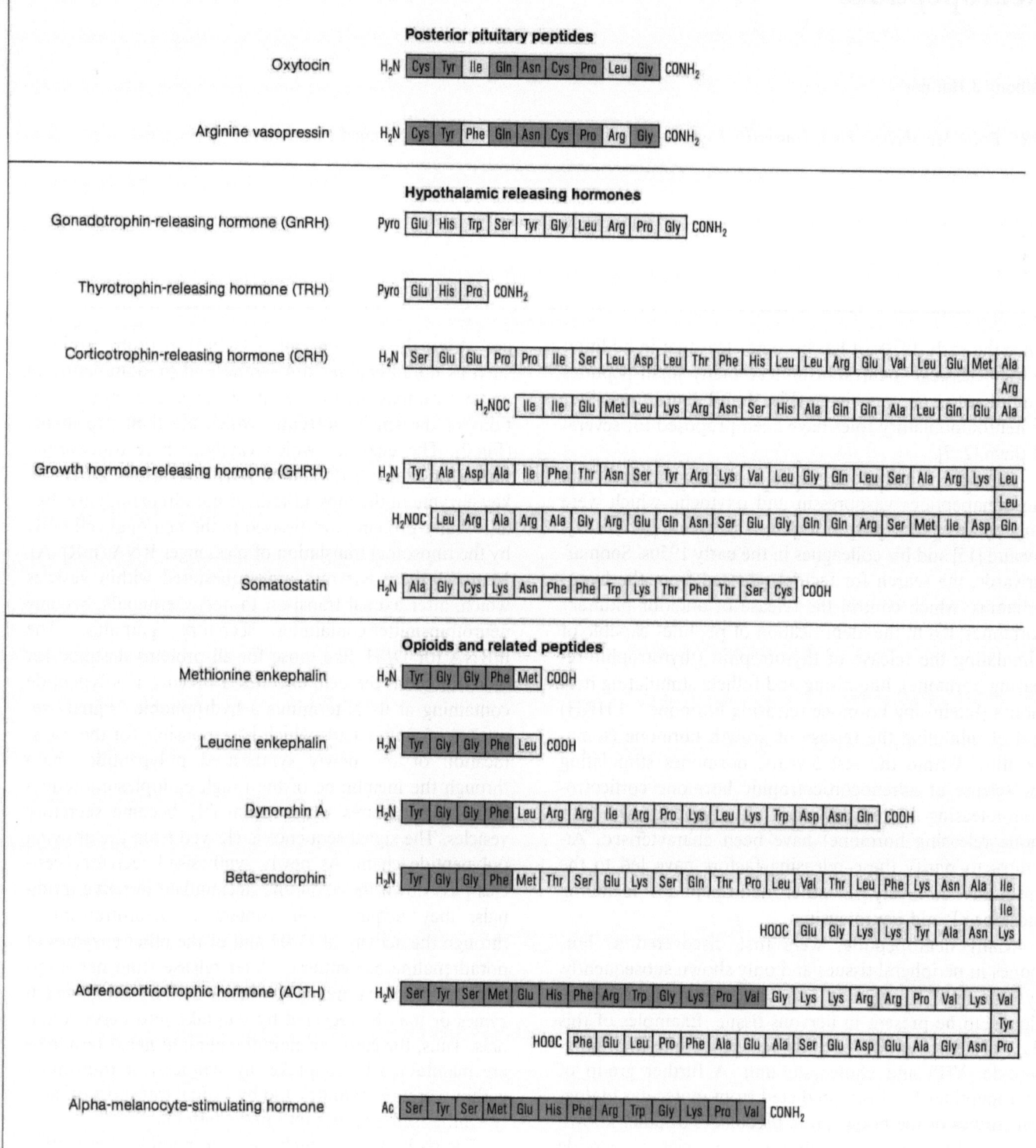

Fig. 1. Amino acid sequences of selected classes of neuropeptides. Amino acid sequences common to more than one member of a neuropeptide family are indicated in **red**

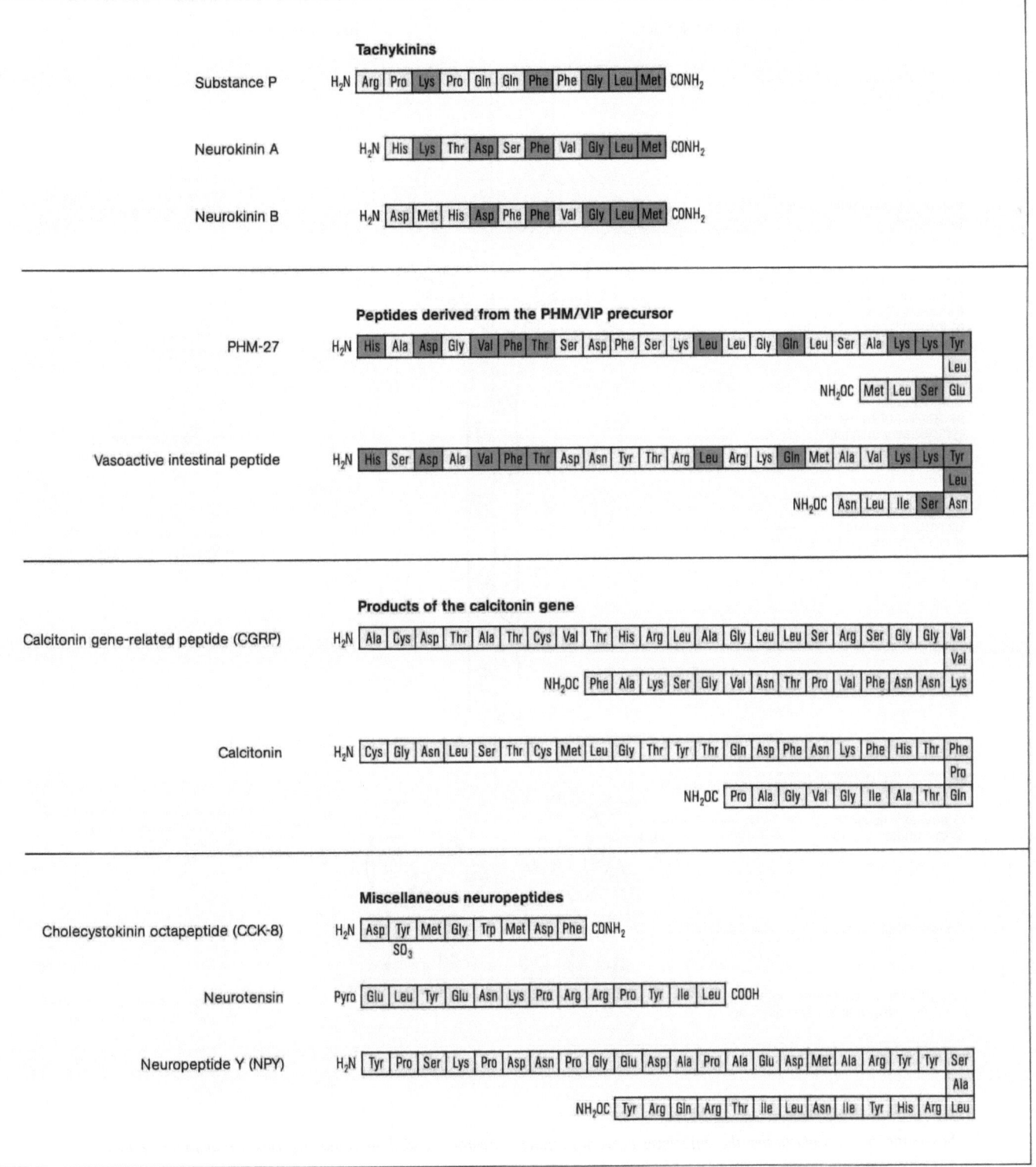

the peptide sequence are known collectively as "post-translational processing" [10]. Whereas terminal stores of monoamine transmitters may be replenished by synthesis and by reuptake into nerve terminals, each peptide molecule released from a nerve ending must be replaced by axonal transport; there is no evidence for the existence of reuptake systems or other mechanisms for the recycling of peptide once released from nerve terminals. This slow and inefficient mechanism should be reflected in the economy with which peptide is utilised in nerve terminals. Peptides may be released intermittently rather than tonically, the amounts stored in terminals may be large rela-

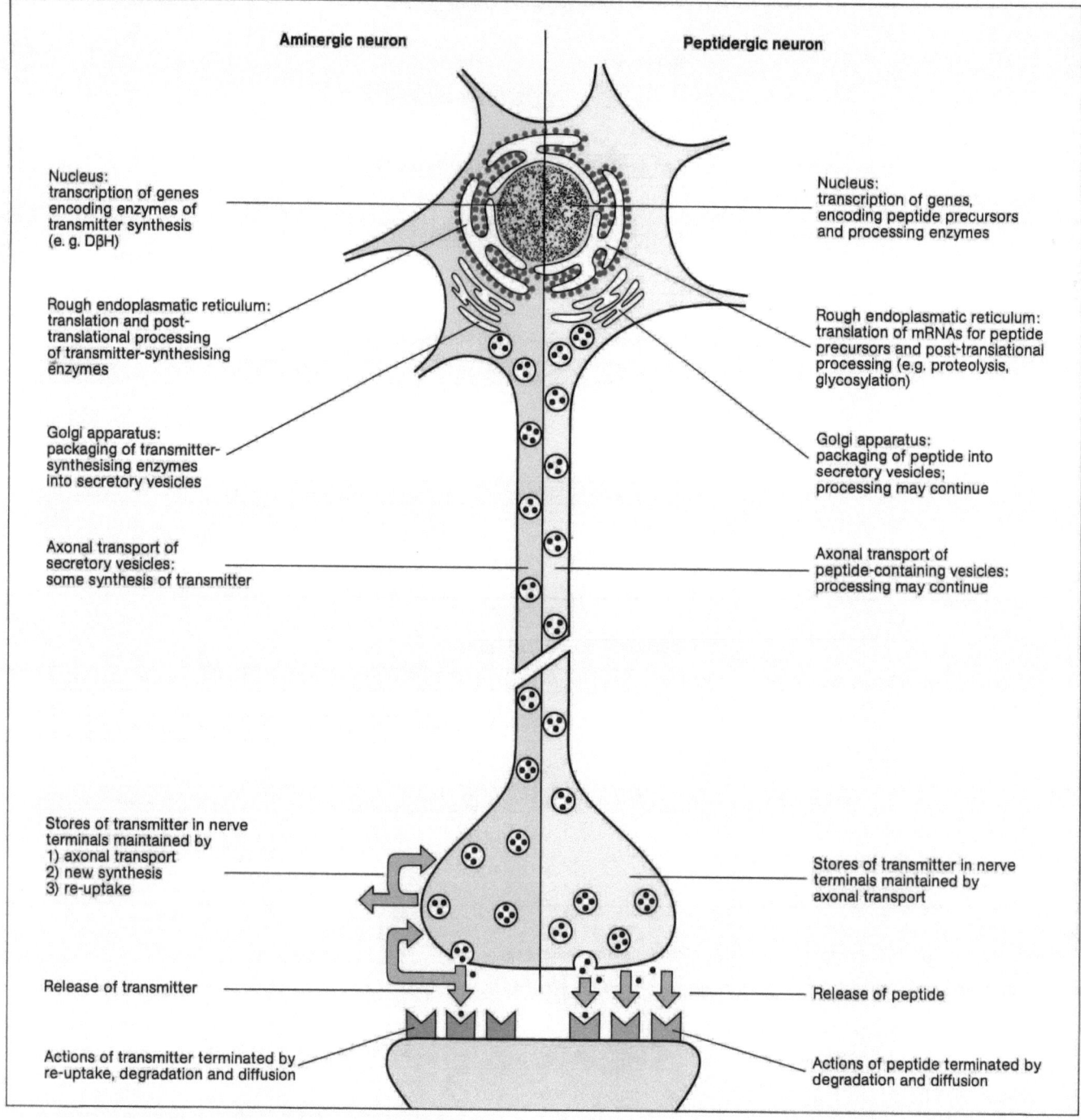

Fig. 2. Schematic diagram illustrating the differences between peptidergic neurones and those containing classical neurotransmitters

tive to the amounts released (i.e. their tunrover may be slow) and the released peptides may activate receptors at much lower concentrations than the "classical" transmitters.

One Neuropeptide Gene: Multiple Products

The amino acid sequences of the precursors to many of the neuropeptides are now known. They are frequently many times larger than the neuropeptide, and usually contain more than one biologically active peptide sequence. In different tissues the same peptide precursor may, as the result of alternative pathways of processing, give rise to different patterns of biologically active peptides. The best-understood example is proopiomelanocortin (POMC; Fig. 3), a polypeptide which, in the anterior pituitary, is processed predominantly to ACTH, β-lipotropin and the endorphin family of opioid peptides. In the intermediate lobe of the pituitary, POMC is processed

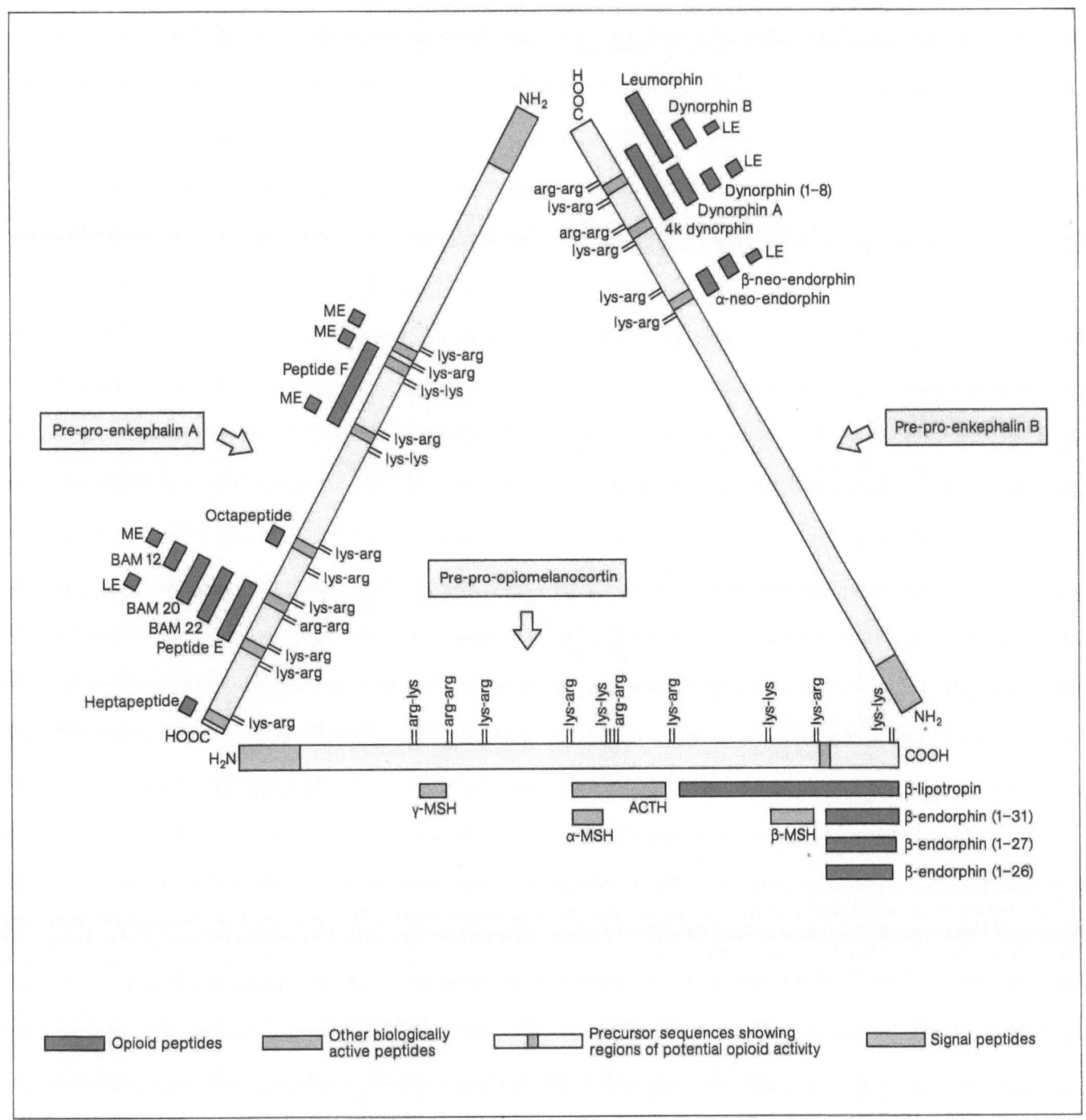

Fig. 3. Schematic representation of the structures of the three human opioid peptide precursors, proopiomelanocortin (POMC), prepro-enkephalin A and prepro-enkephalin B. (Based on [5])

into the melanocyte-stimulating hormones (α-, β- and γ-MSH), a C-terminal fragment of ACTH and biologically inactive forms of β-endorphin [5].

Thus, through variations in post-translational processing, expression of the gene-encoding POMC may give rise in different tissues to alternative patterns of biologically active peptides. A second mechanism by which a single neuropeptide gene may give rise to tissue-specific patterns of peptide products is regulated at the level of the transcription of DNA into RNA.

In eukaryotic cells, the genetic information stored in chromosomal DNA is transcribed into a precursor RNA which is processed in the nucleus into mature mRNA. Most mRNAs are considerably smaller than their nuclear precursors and it is now clear that as an important step in RNA processing, regions of the precursor RNA (introns) are excised and the remaining segments of RNA (exons) are spliced together [1]. Only the RNA associated with the exons appears within the sequence of mature mRNA. In some cases, a single gene may have more than one possi-

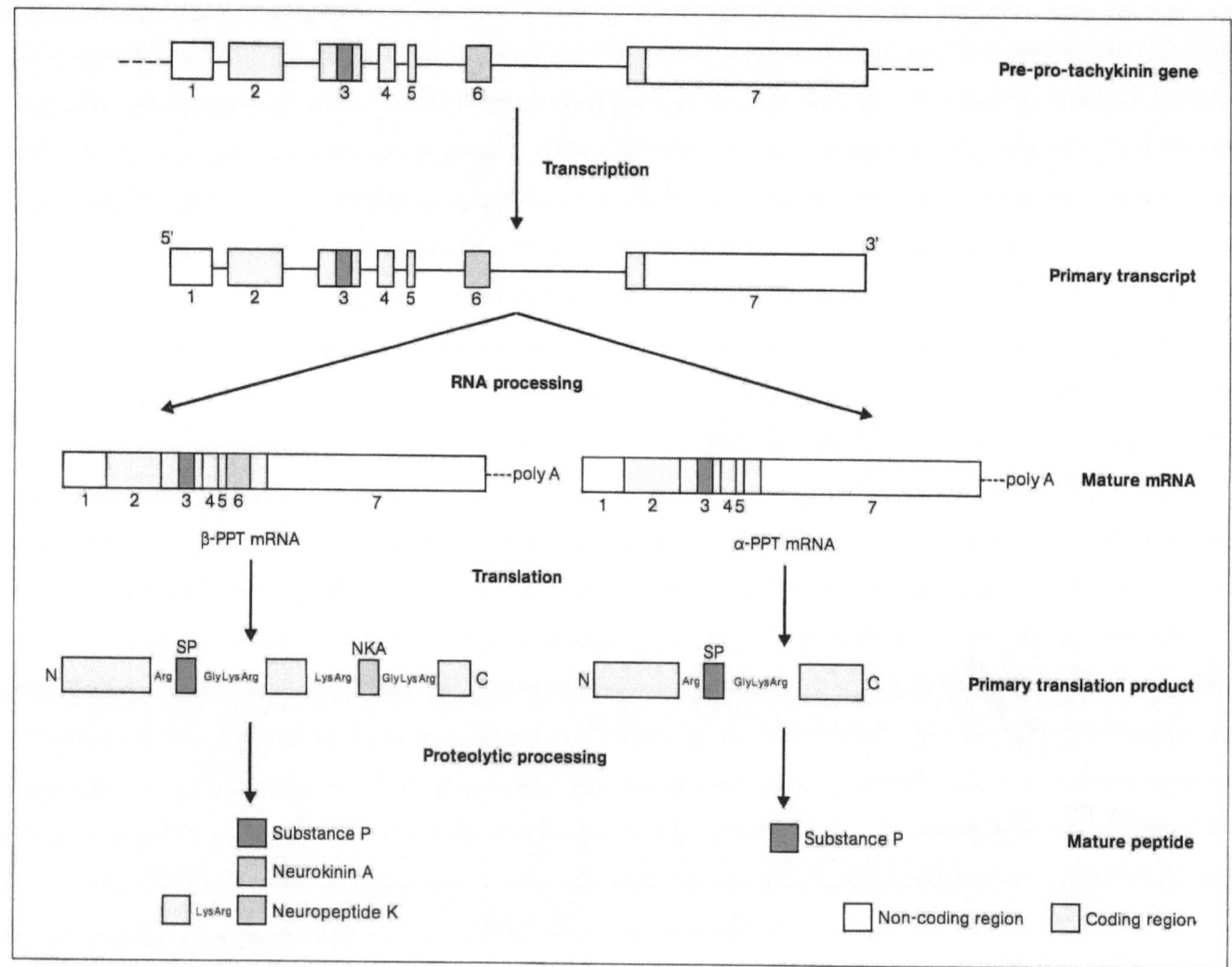

Fig. 4. Generation of two substance P precursor polypeptides from the single preprotachykinin (PPT) gene. The PPT gene consists of seven exons (numbered blocks) separated by six introns. The DNA of the PPT gene is transcribed into a precursor RNA which may be processed in two ways: all seven exons may be spliced together to generate *β*-PPT mRNA, or exon 6 may be omitted to generate *α*-PPT mRNA. *α*- and *β*-PPT are closely similar polypeptides both containing the amino sequence of substance P. *β*-PPT contains the sequence of a second tachykinin, neurokinin A

ble splicing pattern and give rise to more than one species of mRNA.

The first neuropeptide gene to be shown to exhibit variable splicing was that encoding calcitonin, a hormone which plays an important role in the regulation of calcium resorption from bone. The calcitonin gene, which consists of six exons separated by five introns, is expressed in the thyroid gland and in some regions of the nervous system. In thyroid, the first four of the exons are spliced together to form a mRNA coding for the precursor to calcitonin. In brain, mRNA is formed by the splicing together of exons 1-3, 5 and 6 and encodes a precursor to a novel peptide, CGRP, which may play an important role as a neurotransmitter in autonomic and sensory neurones [13].

The preprotachykinin (PPT) gene (Fig. 4), which encodes the precursor to substance P, is also susceptible to differential RNA splicing [11]. The PPT gene consists of seven exons separated by 6 introns. The nuclear precursor RNA corresponding to the PPT gene can be processed in two ways: all seven exons may be spliced together to generate *β*-PPT mRNA, or exon 6 may be omitted to generate *α*-PPT mRNA. *α*- and *β*-PPT are closely similar polypeptides, both containing the amino acid sequence of substance P. However, a sequence of 18 amino acids in the *β*-PPT sequence corresponding to exon 6 of the gene is missing from the *α*-PPT sequence. The missing region contains a sequence of ten amino acids corresponding to a new substance P-like peptide, neurokinin A. Substance P and neurokinin A appear to act through separate classes of tachykinin receptors (so-called SP-P and SP-E receptors), which have different distributions in the nervous system. Thus, through the selectivity of these two tachykinin receptors, the differential expression of two mRNAs from the single PPT gene may ultimately produce different physiological responses.

Opioid Peptides

By 1973 it was clear that both brain and peripheral tissues contained specific receptors for opiates, to which radiolabelled ligands such as dihydromorphine are bound in a saturable and stereospecific manner. It was also known that opiate antagonists such as naloxone could inhibit certain physiological phenomena (for example, the antinociceptive effect caused by electrical stimulation of the periaqueductal grey of the midbrain). These observations suggested that nervous tissue contained endogenous substances with opioid activity. In 1975, Hughes, Kosterlitz and their colleagues reported the structures of the first such "endogenous ligands" for the opioid receptor: Met- and Leu-enkephalin [6]. These two pentapeptides share a common N-terminal amino acid sequence (Tyr-Gly-Gly-Phe-) followed by either a methionine (Met-enkephalin) or a leucine (Leu-enkephalin) residue. Since then, a number of larger neuropeptides containing enkephalin sequences have been described, including the endorphins (containing the Met-enkephalin sequence) and the dynorphins (containing the Leu-enkephalin sequence). There are at least three precursor polypeptides for opioid peptides in brain (Fig. 3; [5]). The endorphins are derived from POMC, the common precursor to ACTH, β-lipotropin and the melanocyte-stimulating hormones. Preproenkephalin A is probably the major source of Met-enkephalin in brain; it contains regions encoding four Met-enkephalin molecules and one Leu-enkephalin and in some tissues may be processed into larger peptides containing the Met-enkephalin sequence. Preproenkephalin B contains three Leu-enkephalin sequences, but the most significant products of its processing may be larger opioid peptides (the dynorphins and neoendorphins).

Each of the three opioid peptide precursors has a characteristic distribution in endocrine tissues and in the nervous system. POMC is synthesised in large amounts in the anterior and intermediate lobes of the pituitary, which release ACTH and endorphin into the bloodstream in response to stress and to a variety of endocrine manipulations. Within the brain, POMC is most abundant in neurones which have their cell bodies in the arcuate nucleus of the hypothalamus and project to the amygdala and to the periaqueductal grey of the midbrain. There is considerable overlap between the distributions of preproenkephalin A and preproenkephalin B, both of which are abundant in areas such as the hippocampus, hypothalamus, nucleus accumbens and spinal cord. However, the adrenal medulla contains large amounts of preproenkephalin A but little or no preproenkephalin B, whereas some regions of the brain (for example the striatum) are rich in peptides derived from preproenkephalin B but appear to contain very little, if any, preproenkephalin A.

Our attempts to understand the functions of the three separate opioid systems in brain have been rendered still more complex by studies of the pharmacology of opioid receptors [16]. There is now strong evidence for the presence in brain of at least three separate subtypes (μ-, δ- and κ) of opioid receptor. Each receptor subtype has a distinctive distribution within the nervous system, which is not strongly correlated with the presence of any particular class of opioid peptide precursor. There are many sites in the brain where opioids have been proposed to function as neurotransmitters or neuromodulators. In any such site, the peptides involved may be derived from any one of three precursors and may act at one or more receptor subtypes; it will clearly be difficult to unravel these complex interactions.

Distribution and Functions of Neuropeptides

It is now clear that a single peptide can, in different tissues, act as a factor acting locally on its cell of origin or on neighbouring cells (modes of intercellular communication termed autocrine and paracrine, respectively) or may fulfil endocrine, neuroendocrine or neurotransmitter functions (Fig. 5). For example, in the pancreas, somatostatin fulfils a paracrine role in the control of glucagon and insulin secretion. Somatostatin synthesised in the intestine is thought to act as an endocrine factor regulating pancreatic hormone release, and somatostatin released from nerve terminals in the median eminence is involved in the neuroendocrine controls of pituitary hormone secretion [8]. Within the brain, pathways containing somatostatin-like immunoreactivity are widespread, and the peptide is likely to function as a neurotransmitter.

Each neuropeptide has been shown to have a distinctive distribution in brain; some regions such as hypothalamus have been shown to be especially rich in a wide variety of neuropeptides, whereas others, such as cerebellum, seem to be almost devoid of neuropeptide-containing neurones. Some neuropeptides (such as somatostatin) are widespread in their distribution, but others (for example LHRH) are restricted to a few cells and a few tracts in the brain.

Little is known of the functions of neuropeptides in the central nervous system (CNS), but in peripheral systems important physiological roles for certain neuropeptides have been proposed. For example, substance P is present in certain sensory neurones which have their cell bodies in dorsal root ganglia. Substance P released from the nerve terminals of these neurones in the spinal cord is thought to be involved in the transmission of painful stimuli [12].

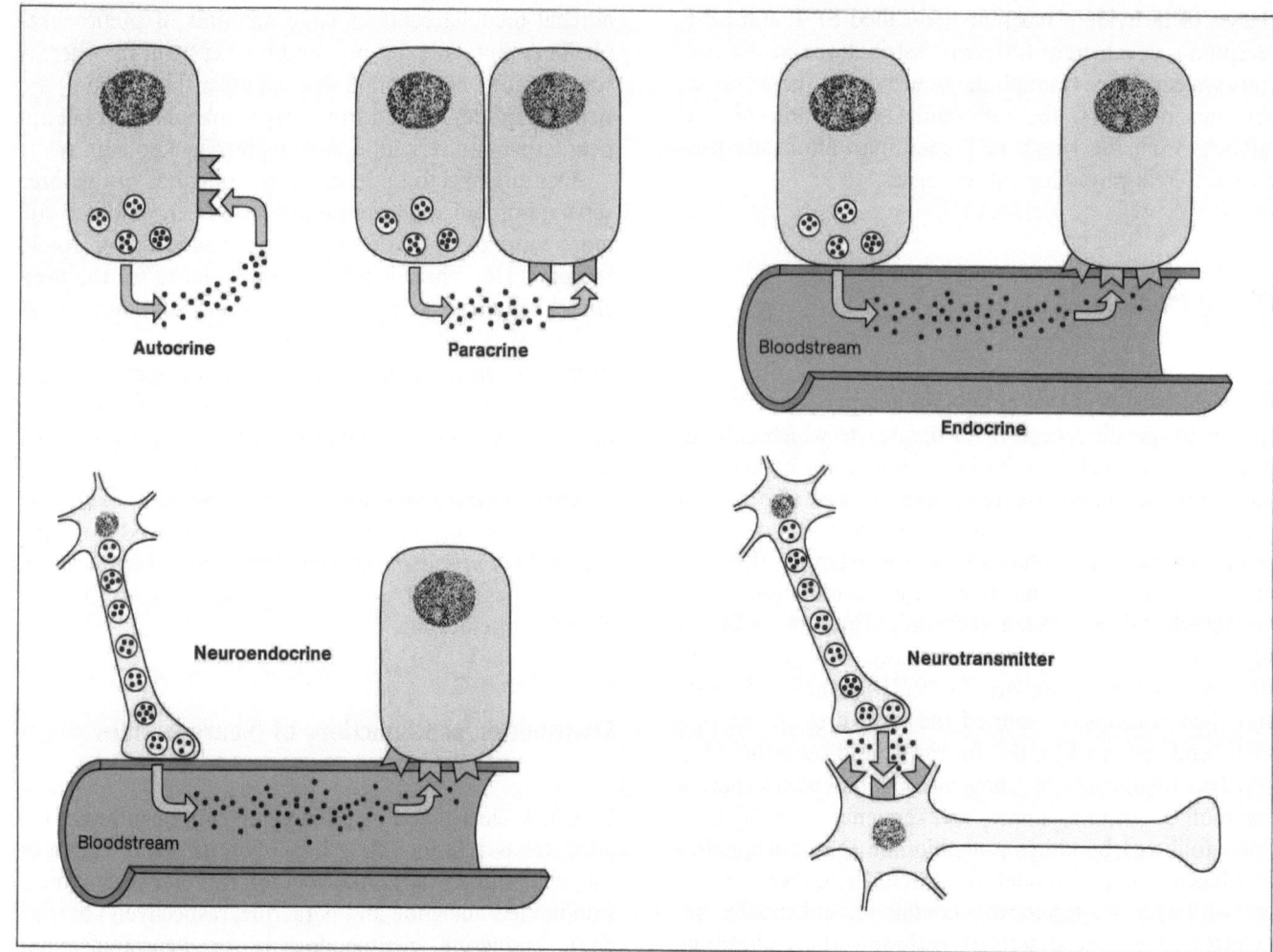

Fig. 5. Functions of peptide-secreting cells. In peripheral tissues, the functions of peptides may be autocrine (acting on the cells from which they are themselves released), paracrine (acting on neighbouring cells) or endocrine (acting on target cells after transport through the blood stream). Within the nervous system, peptides may fulfil a neuroendocrine role after release into the circulation or may act as neurotransmitters

Enkephalin-containing interneurones in the spinal cord may modulate the release of substance P – a possible mechanism for the peripheral actions of opiate analgesics. Substance P released from peripheral endings of the same sensory neurones may regulate vascular permeability and participate in the inflammatory response. In the gut, neurons containing peptides such as VIP and cholecystokinin are abundant and may be important in the regulation of gut motility and in the sensation of visceral pain.

Coexistence of Peptides and Classical Neurotransmitters

There is increasing evidence, derived largely from immunohistochemical studies, that certain neurones containing classical neurotransmitters may also contain one or more of the neuropeptides [9]. For example, in the cerebral cortex, subpopulations of γ-aminobutyric acid (GABA)-containing neurones exhibit cholecystokinin-, somatostatin- and neuropeptide Y(NPY)-like immunoreactivity; there is some overlap between the somatostatin- and NPY-containing subpopulations so that at least some cortical neurones contain three putative neurotransmitter substances. Some neurones in the brainstem contain 5-hydroxytryptamine (5-HT) together with two neuropeptides, substance P and thyrotrophin-releasing hormone. Our understanding of the functions of central neurones containing multiple neurotransmitter candidates is still very limited; however, the study of certain peripheral neurones has shed some light on the problem.

The submandibular gland of the cat is a classical organ for pharmacological studies on autonomic neuroeffector mechanisms. The blood vessels and secretory cells

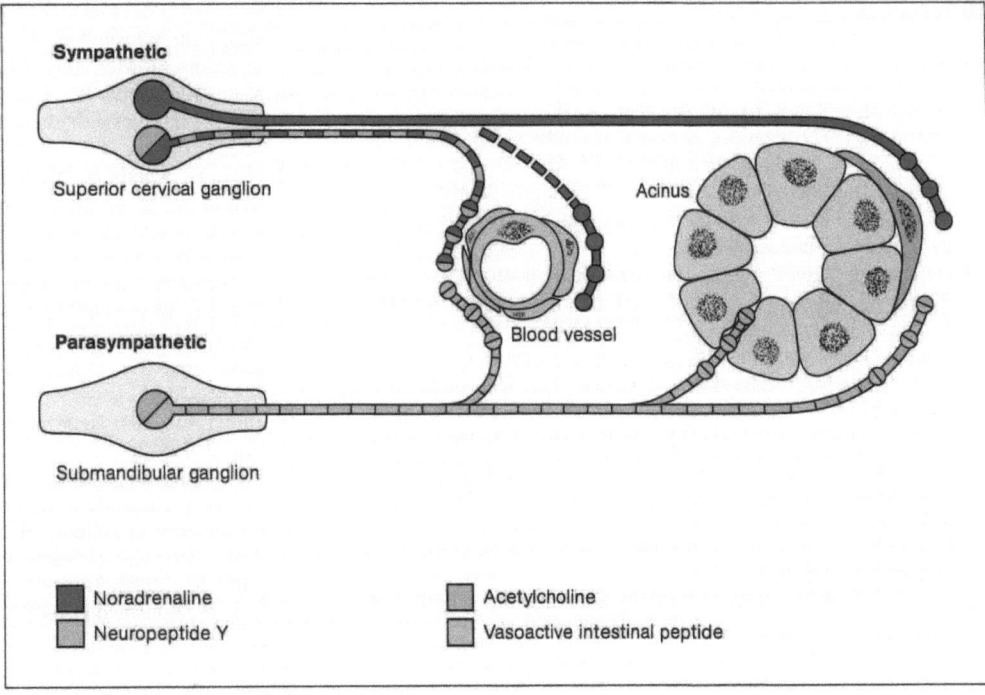

Fig. 6. Innervation of the cat submandibular gland by autonomic neurones containing multiple neurotransmitters. Noradrenaline: induces salivary secretion and vasoconstriction, blocked by α- and β-adrenoceptor antagonists; Neuropeptide Y: no effect on secretion, prolonged vasoconstriction; Acetylcholine: reduces secretion and causes vasodilation, blocked by atropine; Vasoactive intestinal peptide: no effect on secretion, prolonged vasodilation. (Based on [9])

of the gland are innervated by both sympathetic and parasympathetic nerves (Fig. 6). Stimulation of the parasympathetic nerves, which have their cell bodies in the submandibular ganglion, leads to increased salivary secretion and a concomitant increase in blood flow. These effects have long been thought to be due exclusively to the release of the classical neurotransmitter ACh. In a series of studies, Lundberg has shown that the neuropeptide VIP, which coexists with ACh in these nerves, is responsible for a component of the vasodilation observed at high frequencies of nerve stimulation. The sympathetic neurones innervating the submandibular gland have their cell bodies in the superior cervical ganglion and contain the classical neurotransmitter noradrenaline. About half of these cells also contain an NPY-like peptide and these neurones seem to project selectively to blood vessels. Both noradrenaline and NPY are released upon sympathetic nerve stimulation: the release of noradrenaline is thought to bring about an increase in salivary secretion and a strong vasoconstriction of short duration, whilst NPY is probably responsible for the long-lasting vasoconstriction which is observed at high frequencies of stimulation.

The functional significance of the coexistence of neuropeptides with classical neurotransmitters is still not clear. However, in many systems neuropeptides seem to be responsible for a component of the physiological response which is slow in onset and of long duration, whereas the classical transmitters mediate events which are rapid in onset and of short duration. There is also evidence that higher frequencies of stimulation are required for the release of peptide than for the release of the classical transmitter.

Conclusions

Our knowledge of the structures, distribution and biosynthesis of the neuropeptides is now extensive, and the mechanisms by which many of these peptides act on peripheral tissues are now quite well understood. However, for a full understanding of the functions of neuropeptides within the CNS, drugs active as agonists and antagonists at neuropeptide receptors must be developed. Ideally, such analogues should be capable of crossing the blood-brain barrier and thus be centrally active when administered systemically. Research in the last few years has been directed towards the development of novel peptide structures possessing the desired activities; however, drugs of this type have a limited capacity to penetrate the blood-brain barrier. The current explosion of interest in the neuropeptides as potential neurotransmitters or neuromodulators is, to a large extent, due to the discovery that certain plant alkaloids (morphine and its congeners) exert their powerful central and peripheral effects through receptors for which neuropeptides are the endogenous ligands. In the future, it will be important to reverse this deductive process, to identify substances of natural origin which can act as agonists or antagonists at neuropeptide receptors. Such substances may provide the basis for the development of the drugs of the future.

References

1. Darnell JE (1985) RNA. Sci Am 253: 54-64
2. Emson PC (1979) Peptides as neurotransmitter candidates in the mammalian CNS. Prog Neurobiol 13: 61-116
3. Habener JL (1985) Genetic control of hormone formation. In: Wilson JD, Foster DW (eds) Williams textbook of endocrinology. Saunders, Philadelphia, pp 9-32
4. Harmar AJ, Keen P (1981) The turnover of neuropeptides. In: Pycock CJ, Taberner PV (eds) Central neurotransmitter turnover. Croom Helm, London, pp 125-142
5. Hollt V (1985) Multiple endogenous opioid peptides. In: Bousfield D (ed) Neurotransmitters in action. Elsevier, Amsterdam, pp 188-193
6. Hughes J, Smith TW, Kosterlitz HW, Fothergill LA, Morgan BA, Morris HR (1975) Identification of two related pentapeptides from brain with potent opiate agonist activity. Nature 258: 577-579
7. Krieger DT (1983) Brain peptides: what, where, and why? Science 222: 975-985
8. Luft R, Efendic S, Hökfelt T (1978) Somatostatin - both hormone and neurotransmitter? Diabetologia 14: 1-13
9. Lundberg JM, Hökfelt T (1985) Coexistence of peptides and classical neurotransmitters. In: Bousfield D (ed) Neurotransmitters in action. Elsevier, Amsterdam, pp 104-119
10. Mains RE, Eipper BA, Glembotski CC, Dores RM (1983) Strategies for the biosynthesis of bioactive peptides. Trends Neurosci 6: 229-235
11. Nawa H, Kotani H, Nakanishi S (1984) Tissue-specific generation of two preprotachykinin mRNAs from one gene by alternative RNA splicing. Nature 312: 729-734
12. Otsuka M, Konishi S (1985) Substance P - the first peptide neurotransmitter? In: Bousfield D (ed) Neurotransmitters in action. Elsevier, Amsterdam, pp 163-169
13. Rosenfeld MG, Amara SG, Evans RM (1984) Alternative RNA processing: determining neuronal phenotype. Science 225: 1315-1320
14. Turner AJ (1984) Neuropeptide processing enzymes. Trends neurosci 7: 258-260
15. Vigneaud V Du, Lawler HC, Popenoe EA (1953) Enzymic clearage of glycinamide from vasopressin and a proposed structure for this pressor-antidiuretic hormone of the posterior pituitary. J Amer Chem Soc 75: 4880-4881
16. Zukin RS, Zukin SR (1985) The case for multiple opiate receptors. In: Bousfield D (ed) Neurotransmitters in action. Elsevier, Amsterdam, pp 201-208

Methods in the Mapping of Neurotransmitter Systems in the Brain

G. Arbuthnott

MRC Brain Metabolism Unit, University Department of Pharmacology, Edinburgh, United Kingdom

The list of the names of the main telephone exchanges in a country would be of little advantage in understanding the usefulness of the network for its inhabitants. Indeed even the complete description of a telephone handset would not really help in understanding any "functional significance" of the telephone network - though it might suggest what it could not do.

In trying to understand what advantages there are for the animal in having a central nervous system (CNS) we soon hit the same problems. In this chapter we shall try to look at experiments designed to find particular connections in the network of the mammalian CNS. Without them the best chemistry, the most elegant physiology of cell membranes, the most extensive genomic library will not be able to help in answering the simplest functional questions.

First and foremost then, if we want to examine how particular areas of the nervous system are connected, we need to know what a "particular area" is. An area of central nervous tissue where cells show a similar morphology or which is separated from surrounding groups of cells by myelinated fibres is usually referred to as a nucleus. Figure 1 shows an area of hypothalamus where the separation into nuclei is clear. The obvious assumption that all of the cells in a nucleus share a common function, or even common chemistry or identical connections with other parts of the nervous system, is not correct in many cases and leads to further division into "subnuclei". The nervous system has been divided up in many different ways over the centuries, but increasingly the definition of an "area" has become not simply a description of its appearance or position but a list of its connections. Thus although "cortex" just means "outside" or "rind", the motor cortex, originally defined in terms of the behavioural results of stimulation, is better defined as receiving from particular thalamic nuclei and sending axons to other areas concerned ultimately with the motoneurone pool and the skeletal musculature. The major advantage of defining parts of the nervous system by their interconnections (apart from fulfilling the need for names when no behavioural consequences are obvious!) is that it makes homologies between the nervous systems of different animals clearer. The major disadvantage is that the establishing of interconnections is not at all simple.

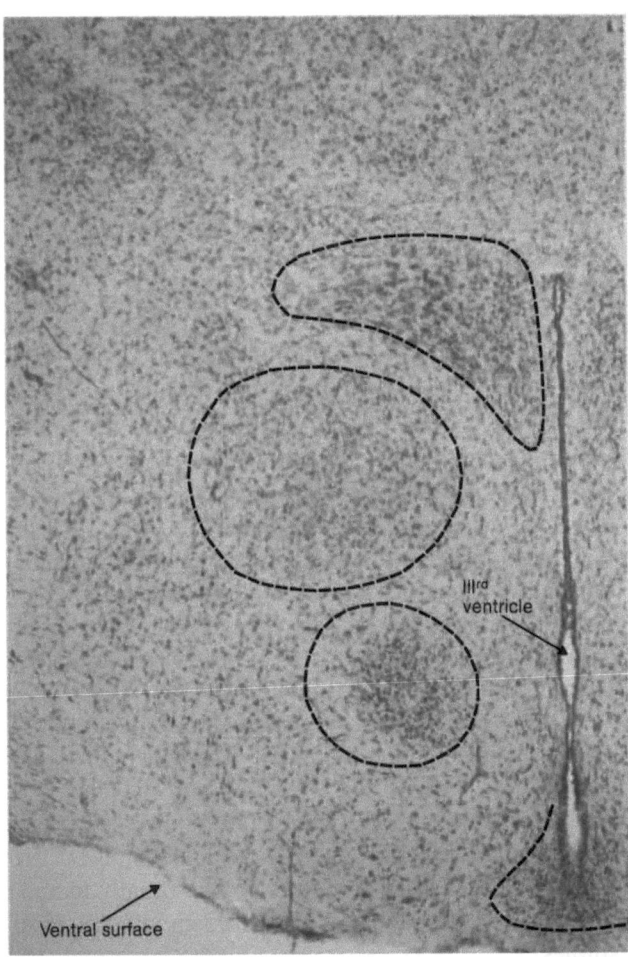

*Fig. 1. Nuclei in the periventricular hypothalamus. This photomicrograph of a section of rat hypothalamus stained with thionin shows several aggregations of neuronal cell bodies. The **dotted lines** indicate the nuclear boundaries visible in this section*

Techniques Depending on Degeneration

Like early "functional studies", and indeed often in close association with them, early attempts at describing con-

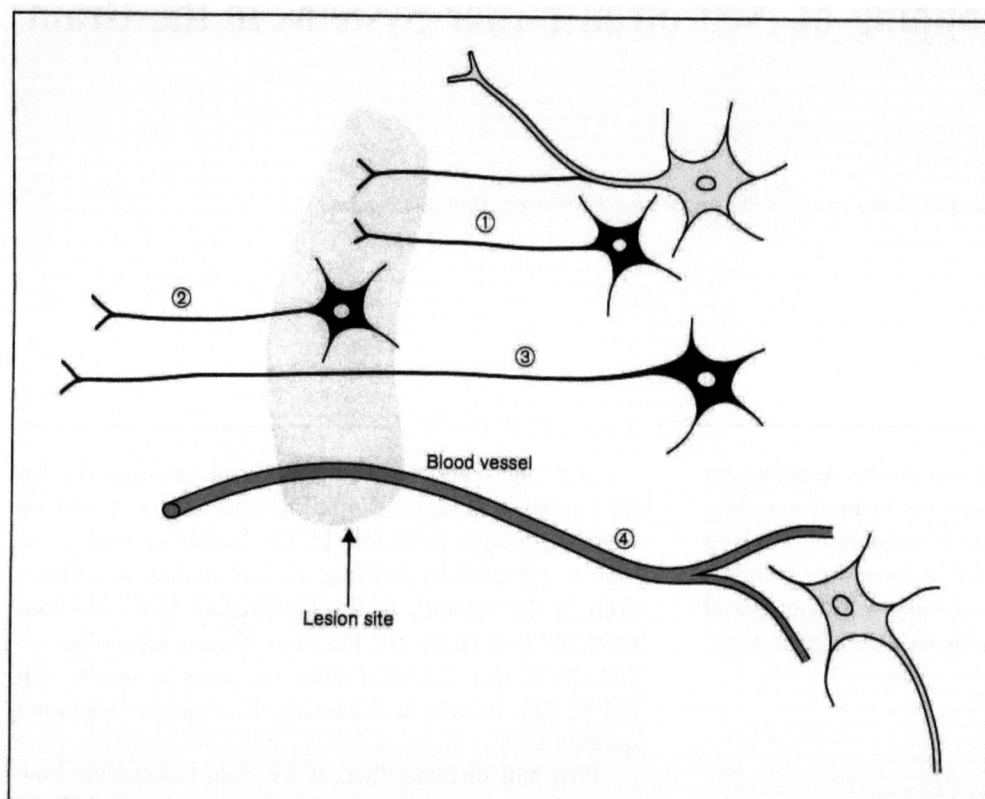

Fig. 2. The different types of degeneration following a lesion in the CNS. The numbers refer to the discussion in the text. *1* Neurones with terminals in the lesioned area die back at least as far as their first branch but often all the way to the cell body. *2* Neuronal cell bodies damaged by the lesion result in anterograde degeneration all the way to the terminals. *3* Axons passing through the lesion result in both anterograde degeneration (as in *2*) and retrograde degeneration (as in *1*). *4* Although blood vessels are damaged by the lesion, little permanent damage is usually done by this to neuronal structure - though function may be affected for some considerable time

nections relied on lesioning and stimulation. Areas of nervous system where gross structure was known could be removed surgically or destroyed electrolytically or by heating. The resultant changes in the behaviour of the animals was studied, and then the anatomical consequences of the lesion examined. A variety of methods are available which allow the description of the pathway taken by degenerating axons that break up as a result of the lesion. Before 1967 only myelinated axons could be reliably followed by these methods. Since then two methods, which also demonstrate the degeneration of unmyelinated fibres have been used [9], though chemically specific methods for the description of anatomical pathways have largely overtaken these. Following degenerating pathways from the lesion site provides simultaneously three kinds of information which need to be separated by other means. (Fig. 2):

1. Neurones whose axons terminate in the lesioned area undergo retrograde degeneration. This degeneration finally results in the total disappearance of the parent cells and so the pathway can be easily recognised. The presence of a branch in the axon, however, can result in degeneration only as far as the branch point. Both the cell body and the undamaged branches survive intact in these cases. Not only is the survival of the cell body misleading, but in some instances of partial damage the intact terminals increase their transmitter content by a so-called pruning effect.

2. Neurones with perikarya destroyed by the lesion will no longer be able to supply the precursors for maintenance of their axons and so these pathways will also degenerate. This anterograde degeneration can usually be followed to the terminal area where the postsynaptic cells will remain largely intact, at least on microscopic examination. Chemical changes and changes in the details of the shape of the target cells have been shown to occur (so-called transneuronal degeneration) but the target neurones do not disappear as a result of the anterograde degeneration of their input.

3. Finally, and causing the maximum of confusion to the investigator, neurones whose axons pass through the lesion site without making any functional connection at all will undergo both the kinds of degeneration just described. To establish beyond reasonable doubt that neurone A proceeds through area B to terminate in C requires many lesions along the whole length of the pathway.

4. Vascular insult in the region of the lesion is a serious complication for functional or behavioural studies. The damage caused to distant structures is usually temporary, and so is not considered a problem in anatomical studies.

In spite of all these problems a final conclusion was often possible and many more recent research reports begin by confessing that the pathway they describe (in a wealth of new detail) was in fact first discovered in the 1890s using methods which depend on degeneration after lesions. Indeed, the earliest proponents of the neurone doctrine, using methods introduced by Golgi and employed so fruitfully by Cajal, described the morphology of individual cells, their axonal pathways and the dazzling diversity of their terminal arborisations in more detail than is available by all but the most revealing of our modern methods.

Studies Which Illustrate Complete Neurones

Golgi's method, which may have been discovered by accident (when a cleaner threw some old, chromate-soaked, central nervous tissue into a bucket containing silver nitrate), suffers from a major disadvantage. The chemical mechanism underlying this method has defied detection. The method is capricious, staining only 10% or so of cells in the most successful cases. The small percentage of cells stained means that each is discernible against a clear background, which would not be true at all if every cell stained, but no control is available, during the experiments, over which cells are stained. Cajal overcame this problem, as have many since, by simply being so prolific and so determined that he made a sufficient number of preparations to have some hope of describing every kind of neurone in a particular area. His comprehensive description of cell morphology performed with such a capricious method is justly famous (Fig. 3). The improvements in the Golgi method have, in general, made it faster (see summary by Ramon-Moliner [23]) or, more recently, applicable to tissue sections [11]. Staining individual sections instead of blocks of tissue makes the Golgi method com-

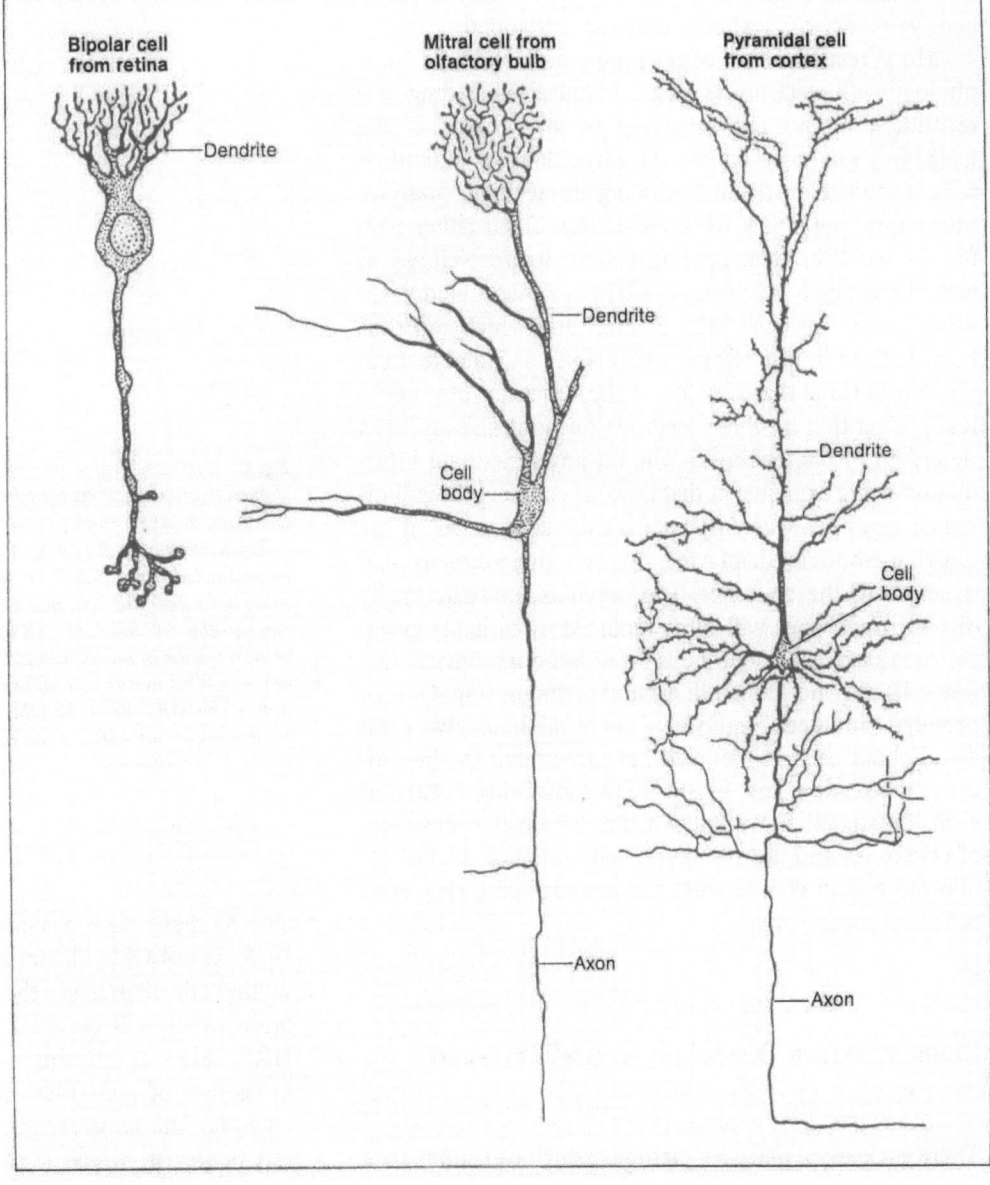

Fig. 3. Differing shapes and sizes of neurones. The cells, stained by the Golgi method, were drawn by Cajal. The bipolar cell is from the retina of a dog, the pyramidal cell from the cortex of a mouse, and the mitral cell from the olfactory bulb (a relay station in the pathway concerned with smell) of a cat

patible with immunocytochemical methods. It is now possible to combine degeneration, Golgi staining and immunohistochemistry in material suitable for study with the electron microscope [26].

Making functional sense of the anatomical detail so described needs other methods. The behavioural consequences of local stimulation (electrical, chemical or mechanical) and lesioning (also electrical, chemical or mechanical) were the first to be studied. Electrophysiological recording techniques are also invaluable in the search for meaningful descriptions of neuroanatomical connections. Because stimulation is also effective on axons of passage, finding the source of particular actions recorded from neurones is often fraught with difficulty. The interpretation of the functional significance of a neuronal network needs both anatomical and electrophysiological expertise. A spectacular success – or rather, a series of successes – has been achieved by the combination of Golgi morphology and electrophysiology. The earlier type of success is exemplified by the very productive collaboration between Eccles and Szentagothai [6] from which a very comprehensive theory of cerebellar function developed.

More recently, this combination of Golgi-type morphology with electrophysiological analysis is leading to a revolution in our understanding of many parts of the CNS. In these experiments, the physiology of individual cells is studied in detail, including intracellular analyses of synaptic potentials. That cell is then filled either with fluorescent dyes (e.g. procoin yellow, lucifer yellow) or with horseradish peroxidase (HRP), a protein whose enzyme activity survives mild fixation and which can thus be visualised in histological sections. It is a suitable compliment to Cajal that very few of the structures physiologically identified by these methods had not already been described in his sections – without any functional information being available at that time, of course. Filling individual neurones with HRP has all the advantages of the Golgi method in displaying the full three-dimensional structure of the neurones. The previous unpredictability of which neurones will fill is replaced by ultimate precision – a particular neurone is filled although only a very few cells may be stained in each experiment (Fig. 4). Like personal "in-depth" interviews, these methods give great detail about each cell studied but are limited by the process of selection and by the time-consuming nature of each investigation. We know a little about the selectivity of electrodes and the addition of physiological identification more than compensates for missing some classes of cells in a chosen area.

Methods Which Depend on Axonal Transport

There is clearly a place for methods which can summarise the inputs and outputs from a given area. The methods which have revolutionised neuroanatomy since 1975 are

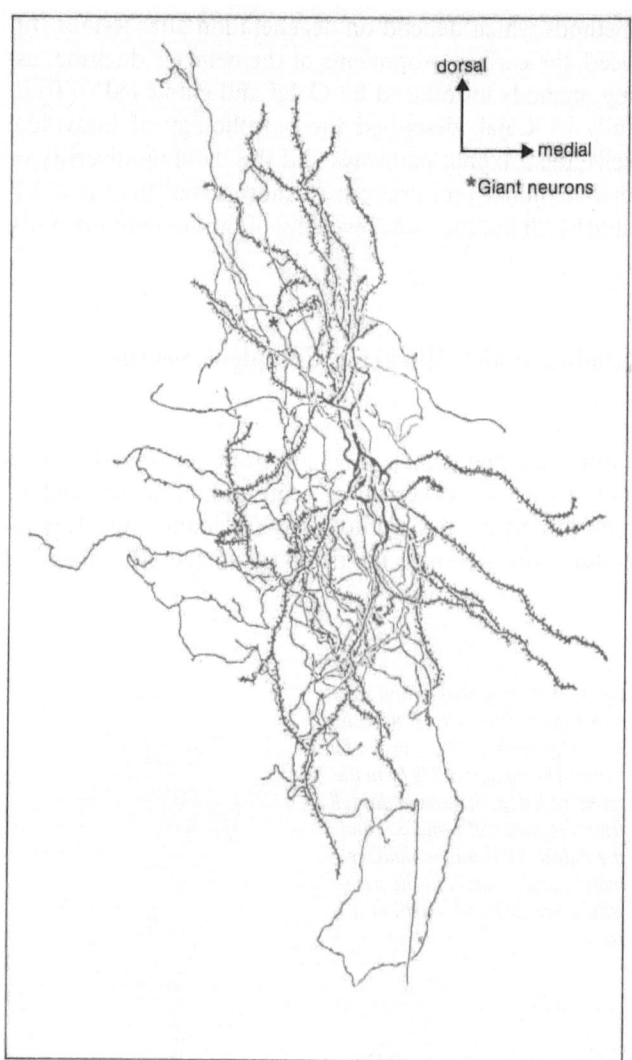

Fig. 4. A striatonigral projection neurone filled with HRP. This densely stained neurone was recorded from intracellularly after being identified as having an axon which projected through the crus cerebri just anterior to substantia nigra. It had an antidromic latency of 7 ms. The drawing reconstructed from seven 80-μ sections of the brain represents the majority of its dendrites and illustrates the recurrent collaterals of its axon close to the cell body. As well as the main axon identified physiologically, a branch of the axon could be followed to the globus pallidus. The cell was filled during recording experiments by MacLeod and Arbuthnott of the MRC Brain Metabolism Unit in Edinburgh, the drawing is by P.Somogyi of MRC Anatomical Neuropharmacology Unit in Oxford

able to make such a summary. At a conference held in 1975, James Olds illustrated the euphoria of the time by saying "all of us have theories about connections in the brain which will eventually be justified by the great god HRP". He was referring not to the single cell method but to the use of peroxidase as a retrograde tracer which is taken up into nerve terminals and transported back to the cell bodies in which it can be visualised by techniques similar to those employed in the single cell studies. It seemed possible that only terminals had the uptake process and that uptake by damaged fibres, if it happened, produced a different pattern of staining in the somata. Neither optimistic assumption generalises easily but certainly intact myelinated fibres take up HRP only weakly and, if damage is limited by the use of small glass injection needles, careful analysis can usually identify terminal areas. More sensitive methods reveal that the tracer moves from cell bodies towards terminals, too. The extra sensitivity in detecting HRP speeds analyses greatly but has the disadvantage that reaction product visualised at the site of the injection covers an area which is very large indeed. The size of injection site - measured as uptake site rather than area covered by reaction product - has been a problem of the method since the beginning. Early experiments suggested really quite small uptake sites, smaller even than the small HRP deposits visible with the original method. Multiple injection sites in cortex resulted in clumps of cells in thalamus in spite of a continuous HRP staining over the injected area [17, 21]. It is still important to check by suitable experiments that uptake from the periphery of the injection site does not distort the results.

Multiple Outputs from One Site

Partly because of the injection site problem, but mainly because of the difficulty of combining HRP with other methods of histochemistry, Kuypers and his colleagues developed methods of tracing which depend on the transport of fluorescent dyes [20]. Except for the problem of autofluorescence induced by glutaraldehyde (which, incidentally, limits the combination of dye tracing with HRP), the dyes are remarkably insensitive to fixation and, since no tissue treatment is necessary for their visualisation, they are easily combined with other chemical or immunological methods of cell identification.

The major advantage of the fluorescence dye tracing is the possibility of combining severel different dye injections in one brain [19].

As noted before, collaterals can sometimes keep the nerve cell body alive in spite of damage to its long axon. It was clear from the methods which visualise axons in sufficient detail that very many neurones do have such collateral branches. By applying several different dyes to proposed terminal areas from particular cell body groups, it is possible to ask the question: do the neurones of the amygdala (for example) project to the contralateral nucleus as well as to the hypothalamus, or are these two terminal areas supplied by different neurones? If single cells contain both dyes then single cells project to both areas. If two populations of cells are contained within the nucleus projecting to the two separate terminal areas then, although both dyes will be found in the nucleus, no single cell will contain both markers. In the real brain, of course, both single- and double-labelled cells are usually found, indicating that a variety of branching patterns are possible within individual "nuclei".

Chemical Specification of Neurons

As histochemistry became a reality rather than a dream, and it became possible to identify the chemical composition of neuronal tissue at the microscopic level, a frustrating disadvantage of the methods rapidly became obvious. From the earliest "Falck-Hillarp fluorescence method", which allowed the visualisation of catechol- and indoleamines in sections of brain [8], to the modern immunochemical techniques [4] all these chemically specific methods show terminal areas and cell bodies clearly but are poor at staining the axonal pathway in between. Some connections were obvious from earlier anatomical work but in general the source of the terminal field in a particular neuronal population has to be demonstrated by lesioning methods. The experiments required to confirm individual pathways have all the troubles associated with the degeneration methods of classical anatomical tracing. Furthermore, the groups of chemically identified cells do not always coincide with conveniently labelled anatomical "nuclei". Nevertheless, the chemistry of the neurones stained is known. It seemed in 1965 that the final proof that amines were transmitters in the CNS had at last arrived [10]. The criteria for identification of a synaptic transmitter contain both items concerned with the presence of a substance in nerve terminals and their synthesis and degradation in the area, and also items concerned with showing that the action of the suggested substance is identical to the effect of stimulation of the presynaptic terminal. The new "chemical neuroanatomy" has had the effect of increasing the emphasis on the criteria concerned with "presence" and decreasing the emphasis on the "identity of action" conditions. Thus we have more than 30 peptide "neurotransmitters" [16]. We have evidence for the coexistence of peptides and "classical" (ever since 1965) neurotransmitters. Dale's "principle" (as stated by Eccles [5]) "that individual neurones may secrete the same substance at each of their terminals" has come under attack. Nevertheless, still in 1986 we had trouble with Paton's identity of action [22] criteria for a transmitter substance in the case of amines (let alone peptides) although no-one doubts the "site of synthesis and degradation" criteria at all.

Functional considerations aside, the brain is rapidly being compartmentalised in ways not accessible to Golgi and Cajal. Pathways not described, or still the subject of debate, are now obvious because the neurones involved all contain the same chemical marker. The widespread connections of locus coeruleus [18] (literally the "blue" place) would never have been guessed by methods which did not rely on the fact that the cells contain noradrenaline and so can be proved to be the source of noradrenaline-containing terminals in thalamus, cortex, cerebellum and spinal cord.

Neurotoxins

For amines, though less so – thus far – for other transmitter candidates, the neuroanatomical structures which they form, and more importantly perhaps the behavioural and neurophysiological consequences of their loss, have been much easier to determine since the discovery of "specific" neurotoxins [1, 28]. 6-Hydroxydopamine (6OHDA) and 5,7-dihydroxytryptophan (5,7-DHT) are specific because of the selectivity of the uptake systems in neurones which concentrate them. Nevertheless, by a collection of sophisticated pharmacological and surgical tricks, they can be made to destroy only those neurones which possess the relevant uptake system. The resultant damage to the animal's behavioural repertoire may well suggest the symptoms of human diseases (e.g. the damage to dopamine might suggest Parkinson's disease) but these are excellent tools for neuroanatomical use. Only 5-hydroxytryptamine (5-HT)-containing cells in the brainstem die after 5,7-DHT applied locally. Substance P as well as 5-HT concentrations in spinal cord are reduced by such local application of toxin [14]. It thus seems likely that 5-HT and substance P coexist in the descending neuronal systems destroyed by the toxin. Which is the "neurotransmitter", which the "modulator"? Are both released by the same stimuli? Which subserves which "postsynaptic action"? So far no clear behavioural consequence of the loss is described – do these systems have a unique "function" in the adult animal? The functional questions remain, but the anatomy is the more certain for the demonstration of the damage to substance P-containing cells as a result of the application of a toxin whose action depends on the ability of the cells to concentrate 5-HT.

Immunohistochemistry

The Falck-Hillarp method depends on the specific chemical reaction between the monoamines and formaldehyde, but the demonstration of peptide-containing neurones depends on the more generally applicable antibody-antigen detection system. Antibodies to small molecules complexed to a larger protein or "hapten" have become the major tool with which to delineate the chemical characteristics of groups of neurones. Monoclonal antibodies raised against unknown determinants in a "simple" nervous system, like that of the leech, pick out particular groups of neurones. Antibodies raised to 5-HT provide permanent and greatly more detailed preparations with which to describe the anatomy of the 5-HT-containing

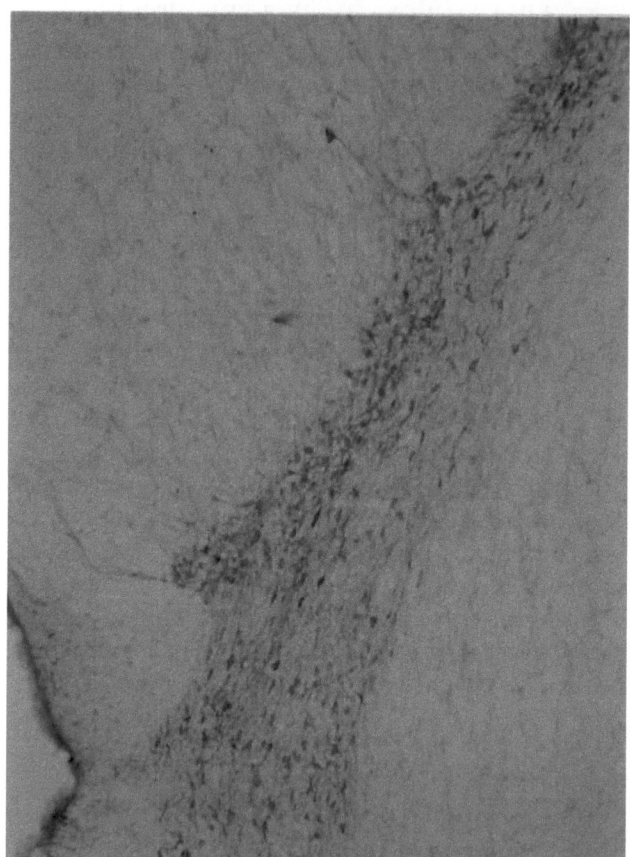

Fig. 5. Dopamine-containing cells in the substantia nigra. These neurons are stained by an antibody method detecting tyrosine hydroxylase. Since these cells do not stain with antibodies raised against dopamine β-oxidase, it is likely that they make dopamine but do not continue the synthesis to noradrenaline

systems in brain. Antibodies have been prepared against all the major synthetic enzymes which synthesise the catecholamines; that is, tyrosine hydroxylase (TH; which catalyses the conversion of tyrosine to dopa); dopamine-β-oxidase (DβO; which catalyses the formation of noradrenaline from dopamine); and phenylethanolamine-N-methyltransferase (PNMT; which catalyses the conversion of noradrenaline to adrenaline). These three enzymes in particular have characterised, in ways not really accessible to the "pharmacological" methods of the 1960s, the source and distribution of individual systems of adrenaline-, noradrenaline-, and dopamine-containing neurones in brain. Neurones with TH but no DβO clearly cannot synthesise noradrenaline and so must be dopamine-containing (Fig. 5). Those with PNMT immunoreactivity are likely to make adrenaline in preference to noradrenaline, which is the likely transmitter in cells containing TH and DβO but no PNMT.

Antibodies are not only raised against enzymes, but many molecules relevant to our understanding of neurobiology, transmitter candidates, nerve growth factors and

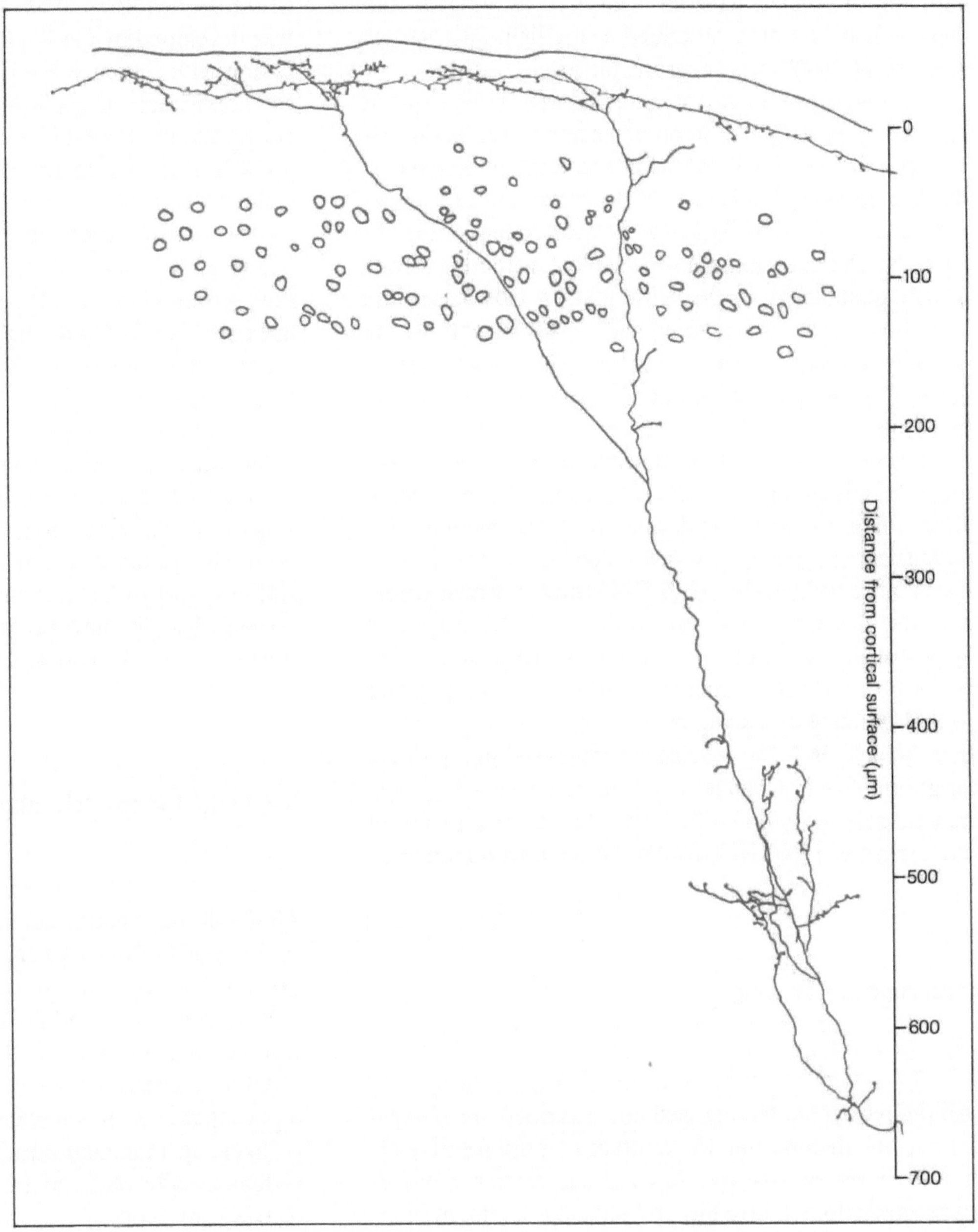

Fig. 6. *Terminals of cells in the ventromedial nucleus of the thalamus ramify in layer I of the cerebral cortex. This camera lucida drawing of individual terminal distributions comes from the collection of neurones filled by an injection of PHA-L in the ventromedial thalamus. There seem to be two distinct types of morphology, both of which are represented in this figure. The ability to see such fine detail in the area innervated is the major advantage of the PHA-L method*

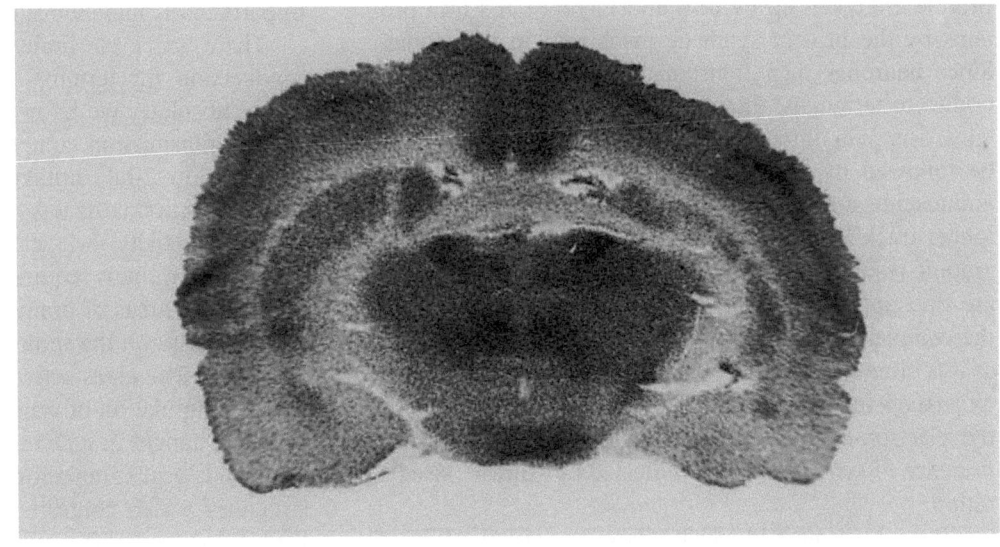

Fig. 7. *2-DG autoradiograph showing increased uptake in layer IV of cortex. This autoradiographic section through a rat brain taken from a normal control rat shows areas of higher activity as **darker areas** on the section. Layer IV of cortex is particularly clear as are several thalamic areas*

adhesion molecules, have all been used as antigens and then their distribution described at the light-microscopic and also at the electron-microscopic level.

The most careful investigators usually refer to what has been seen as X-like immunoreactivity. The main reason for this is that it is impossible to exclude the chance that the antibody binds to a small component of a larger molecule which has a sequence identical to that of the antigen (X). Present immunohistochemical methods are only grossly quantitative in the sense that, in most cases, the immunoreactivity is either present or absent and, in the majority of cases, increases in antigen concentration must result in increased numbers of positive cells if they are to be visible.

Recently it has become possible to ask which structures in microscopical sections possess the messenger RNA for a particular molecule. Now the methods for identifying the messenger RNA depend on the tissue's own ability to hybridise with DNA probes which recognise the RNA required for synthesis of the substance against which the probe was generated. These in situ hybridisation methods, which rely on autoradiography, are in their infancy as tracing methods but provide evidence that the antigen being studied is synthesised in a particular group of cells. Changes in the amount of RNA present may be able to be quantified, and thus some measure of the control of DNA transcription in the cells obtained.

Anterograde Tracing

All the retrograde tracing methods described are in regular use but all carry the disadvantage that uptake may occur in fibres of passage. Anterograde tracing methods exist which do not have this problem, at least not in intact animals - there is some doubt about isolated nerve fibres in vitro [7]. Tritiated (^{3}H) amino acids such as leucine or proline are taken up by cells and incorporated into proteins by the protein synthetic machinery in cell bodies. Since neurones then transport these proteins from cell bodies to terminals, they also move the radioactive proteins. The path of the axons to their terminations can thus be followed by autoradiography. Although ^{14}C-labelled aminoacids are also incorporated and transported, the longer track length of the ^{14}C leads to worse autoradiographic resolution but to shorter exposure times. Proteins are transported at several rates along the axons, and so the complete pathway is visible. This has some problems as it is often difficult to be sure that there are no synapses *en passage* in certain areas. The technique can be used at the electron-microscopic level, however, and there the presence of synapses can be indicated by "direct" visualisation.

Recently, a method which allows the visualisation of terminal areas, indeed of "Golgi-like" filling of whole neurones, dendrites, axons and terminals, as well as soma, was developed by Gerfen [13] and used by him in combination with ^{3}H-leucine tracing [12]. The method requires the application of a plant lectin (*Phaseolus vulgaris* leucoagglutinin - PHA-L) locally to the brain by iontophoresis. Close to the injection site, some 100 neurones are stained when the tissue is reacted with an antibody to the lectin and then processed by immunohistochemical methods (Fig. 6). The axons of these cells can be followed to their terminations, and the morphology of individual synaptic sites studied in the light microscope. Again, the only improvement on Golgi is in *choosing* which cells are filled. Unlike the single cell filling with HRP, no sampling bias due to the electrode is seen; all the cells in the region of the injection concentrate the tracer. The physiology of the injected cells is not demonstrated as directly as with single cell filling but about 100 cells per experiment can be studied rather than one or two. The PHA-L method is still new and so has not been applied in many areas yet. Nevertheless, it may be an excellent choice for future tracing of efferent pathways from groups of cells.

Regional Energy Metabolism

Although they should not really be regarded as methods for the study of detailed anatomical synaptic connections, the developments in the measurement of regional metabolic rate are rapid and the potential enormous. The basic method depends on the discovery that 2-deoxyglucose (2-DG) is converted in nerve cells to 2-deoxyglucose-6-phosphate which is metabolised only slowly. The 2-DG is taken up into neurones by the active glucose uptake system usually required by the neurone to maintain its metabolism. Since glucose utilisation by neurones is necessary for energy metabolism, and indeed is stoichometrically related to O_2 upake, the method allows the detection of particularly metabolically active regions in brain [25].

There are clear indications that the method has a predilection for terminal areas although this is by no means absolute. We know little about the energy costs of axonal transmission compared to transmitter release, but the very low 2-DG uptake in myelinated tracts suggests that conduction costs less than release. Similarly, the band of higher activity in layer IV of the cerebral cortex suggests that terminals require more glucose than cell bodies (Fig. 7). The areas of brain responding to stimulation of a single vibrissa on the snout of a rat have been outlined by 2-DG [15]. The areas activated by pharmacological agents or by manipulation of water intake [27] have also been reported. Changes in 2-DG uptake also follow lesions in the CNS and local stimulation but the uncertainty about the structures which contribute to the increased metabolism leads to obvious difficulties in interpretation.

The full potential of autoradiographic mapping is

seen in the recent development of positron emission tomography. Positrons are emitted by some unstable isotopes of common elements which can be generated in a cyclotron. Each emission gives rise to a pair of γ rays each with identical energy and exactly 180° apart in space. Detection of the angle of emission and the incident energy of a pair of rays defines the position of the positron emitter in space. The shape of the structure which contains the emitter can then be determined, but more importantly, the dynamic concentration of the substance can be followed in sequences of tomographic images.

Early results from this method, which is not suitable for small animals at present (the resolution is approximately 1 cm^3), include estimates of the rate of metabolism of *L*-dopa in human striatum [3]; changes in fluorinated 2-DG metabolism in Alzheimer cortex [2], and, surprisingly, the description of an area in the parahippocampal gyrus where cerebral blood flow is asymmetric in patients with a panic-anxiety syndrome [24]. Clearly, following the action through areas of human brain like this opens a new chapter in the description of functional neuroanatomy.

References

1. Baumgarten HG, Björklund A, Lachenmayer L, Nobin A (1973) Evaluation of the effects of 5,7-dihydroxytryptamine on serotonin and catecholamine neurons in the rat CNS. Acta Physiol Scand [Suppl] 373:1-15
2. Benson DF, Mettler EJ, Kuhl DE, Phelps ME (1982) The application of positron emission computed tomography in neurobehavioural problems. In: Kertesz A (ed) Localization in neuropsychiatry. Academic, New York
3. Calne DB, Langston JW, Martin WRW, Stoessl AJ, Ruth TJ, Adam MJ, Pate BD, Schulzer M (1985) Positron emission tomography after MPTP: observations relating to the cause of Parkinson's disease. Nature 317: 246-248
4. Cuello AC (1983) Immunohistochemistry. IBRO handbook series: Methods in the neurosciences 3. Wiley, Chichester
5. Eccles JC, Fatt P, Koketsu K (1954) Cholinergic and inhibitory synapses in a pathway from motor axon collaterals to motorneurons. J Physiol (Lord) 126: 524-562
6. Eccles JC, Ito M, Szentagothai J (1967) The cerebellum as a neuronal machine. Springer, Berlin Heidelberg New York
7. Edström A, Hanson M (1973) Temperature effects on fast axonal transport of proteins 'in vitro' in frog sciatic nerves. Brain Res 58: 345-354
8. Falck B, Hillarp N-A, Theime G, Torp A (1962) Fluorescence of catecholamines and related compounds condensed with formaldehyde. J Histochem Cytochem 10: 348-354
9. Fink RP, Heimer L (1967) Two methods for selective impregnation of degenerating axons and their synaptic endings in the central nervous system. Brain Res 4: 369-374
10. Fuxe K (1965) Evidence for the existence of monoamine neurons in the central nervous system. IV Distribution of monoamine terminals in the central nervous system. Acta Physiol Scand 64 [Suppl] 247: 36-85
11. Gabbott PL, Somogyi J (1984) The 'single' section Golgi-impregnation procedure: methodological description. J Neurosci Methods 11: 221-230
12. Gerfen CR (1985) The neostriatal mosaic. 1. Compartmental organisation of projections from the striatum to the substantia nigra in the rat. J Comp Neurol 236: 454-476
13. Gerfen CR, Sawchenko PE (1984) An anterograde tracing method that shows the detailed morphology of neurons, their axons and terminals: immunohistochemical localization of an axonally transported plant lectin, *Phaseolus vulgaris* leucoagglutinin (PHA-L). Brain Res 290: 219-238
14. Gilbert RFT, Emson PC, Hunt SP, Bennett GW, Marsden CA, Sandberg BEB, Steinbusch HWM, Verhofstad AAJ (1982) The effects of monoamine neurotoxins on peptides in the rat spinal cord. Neuroscience 7: 69-88
15. Gonzalez MF, Sharp FR (1985) Vibrassae tactile stimulation: (^{14}C)2-deoxyglucose uptake in rat brain stem, thalamus and cortex. J Comp Neurol 231: 457-472
16. Hökfelt T, Johansson O, Goldstein M (1984) Chemical anatomy of the brain. Science 225: 1326-1334
17. Jones EG (1975) Possible determinants of the degree of retrograde neuronal labeling with horseradish peroxidase. Brain Res 85: 249-253
18. Jones BE, Yang T-Z (1985) The efferent projections from the reticular formation and the locus coeruleus studied by anterograde and retrograde axonal transport in the rat. J Comp Neurol 242:56-92
19. Kuypers HGJM, Bentivoglio M, Catsman-Berrevoets CE, Bharos AT (1980) Double retrograde neuronal labeling through divergent axon collaterals, using two fluorescent tracers with the same excitation wavelength which label different features of the cell. Exp Brain Res 40: 383-392
20. Kuypers HGJM, Catsman-Berrevoets CE, Padt RE (1977) Retrograde axonal transport of fluorescent substances in the rat's forebrain. Neurosci Lett 6: 127-135
21. Mesulam MM (1982) Tracing neuronal connections with horseradish peroxidase. IBRO handbook series: Methods in the neurosciences, 1. Wiley, Chichester
22. Paton WDM (1958) Central and synaptic transmission in the nervous system (pharmacological aspects). Ann Rev Physiol 20: 431-470
23. Ramon-Moliner E (1970) The Golgi-Cox technique. In: Nauta WJH, Ebbesson SOE (eds) Contemporary research methods in neuroanatomy. Springer, Berlin Heidelberg New York, pp 32-55
24. Reiman EM, Raichle ME, Butler FK, Herscovitch P, Robins E (1984) A focal brain abnormality in panic disorder, a severe form of anxiety. Nature 310: 683-685
25. Sokoloff L (1985) Basic principles in imaging of regional cerebral metabolic rates. In: Sokoloff L (ed) Brain imaging and brain function. Raven, New York, pp 21-49
26. Somogyi P, Freund TF, Wu J-Y, Smith AD (1983) The section Golgi impregnation procedure. 2 Immunocytochemical demonstration of glutamate decarboxylase in Golgi impregnated neurons and in their afferent synaptic boutons in the visual cortex of the rat. Neuroscience 9: 475-490
27. Sutherland RC, Martin MJ, McQueen JK, Fink G (1983) Water deprivation results in increased 2-deoxyglucose uptake by the paraventricular neurones as well as pars nervosa in Wistar and Brattleboro rats. Brain Res 271: 101-108
28. Ungerstedt U (1968) 6-Hydroxydopamine induced degeneration of central monoamine neurones. Eur J Pharmacol 5: 107-110

Molecular Aspects of Central Neurotransmitter Function

Rory Mitchell

MRC Brain Metabolism Unit, University of Edinburgh, Department of Pharmacology, Edinburgh, United Kingdom

Introduction

It is clear from the previous chapters that the list of substances that may be considered as possible neurotransmitters in the central nervous system is rapidly growing. In addition to the now 'classical' neurotransmitters such as acetylcholine, monoamines and amino acids (amongst which there are also new tentative candidates such as adrenaline and taurine), there are now scores of neuropeptide candidates. Few of these have been shown to fully satisfy the criteria for acceptance as neurotransmitters. Nevertheless, in many cases there is evidence fully consistent with, and suggestive of, a role in chemical neurotransmission. A general impression has developed that neural actions of peptides are inordinately slow in onset and offset and that they should therefore be regarded more as long-term neuromodulators than as true transmitters. This may relate in part to our limited ability to rapidly deliver adequate concentrations of peptides to the correct neuronal loci, due to factors such as poor ejection from micropipettes and the presence of powerful peptide-degrading enzymes in neuronal tissue [31]. Indeed, in some instances, for example, in the excitatory action of cholecystokinin (CCK8) on CAI pyramidal cells in rat hippocampal slices [31], peptides can act at least as rapidly as the classical excitant *l*-glutamate. In other cases (perhaps more usually), peptide actions in the nervous system do take place over a somewhat slower time course, although they are still of great relevance to neurotransmission. Examples include the late slow depolarisation elicted by an LHRH-like peptide in bullfrog sympathetic ganglia [27] and the slow depolarisation in guinea-pig inferior mesenteric ganglia mediated by substance P [77]. This is not unique to peptides, however; for example, slow depolarisations of frog sympathetic neurones are also observed in response to acetylcholine [13]. Indeed, there are many recorded examples of slow postsynaptic potentials elicited by 'classical' neurotransmitters at a variety of central or peripheral sites [23].

Any substance secreted from a neurone in response to a specific stimulus may contribute to information transfer between that and another cell, providing that it can be detected and that in some way a signal can be interpreted. Experimental evidence increasingly indicates that the nervous system operates very differently from a simple balance between discrete excitatory and inhibitory inputs. Such a system would have to be essentially a hard-wired layout with point-to-point inputs to every cell that required influence, even given the subtlety conferred by summation or occlusion of different inputs. The huge diversity of chemical signals suggests instead that there must be a spectrum of classes of information carried [26], and this is supported by the great variety of influences observed in postsynaptic membrane properties [23, 64]. Many neuropeptides, as well as monoamines, are released from terminals which are not in tight synaptic contact with other cells and may therefore diffuse over a broader range, influencing an array of receptive cells to a greater or lesser degree. There can thus be a broad sphere of influence without point-to-point contact (see for example [27]) and considerable scope for plasticity in influence. Localisation of receptive sites on the dendrites or body of neurones may also contribute to the selective gating of inputs to particular zones of the dendritic field. Many of the available signals now appear to represent changes in excitability rather than on or off switching. In contrast to nicotinic cholinergic receptors (clearly characterised at the neuromuscular junction), muscarinic cholinergic receptors are now known to mediate slow and long-lasting increases in excitability of neurones [13], probably by inhibition of a particular membrane ion channel that selectively allows K^+ to leave the cell. Opioid peptides (enkephalins) and drugs like morphine hyperpolarise and depress the firing rate of locus coeruleus neurones also by increasing the cell membrane conductance to K^+ [81]. A variety of neurotransmitters can reduce the duration of action potentials in chick dorsal root ganglion neurones, apparently by inhibition of membrane Ca^{2+} conductance [19]. In *Helix* neurones, dopamine can decrease the amplitude of the spike plateau by inhibiting Ca^{2+} conductance or can increase it by blocking an outward K^+ conductance [53]. Even the adaptation to a series of depolarising stimuli that are dependent on a Ca^{2+}-activated K^+ conductance can be inhibited by noradrenaline [35]. Thus, in terms of membrane electrical properties, many levels of modification can be produced. It is clear, though, that neurotransmitters can elicit a range of metabolic as well as electrical changes in cells. The production of intracellular second messengers in response to neurotransmitters and hormones has been intensively studied, particularly in relation to the critical control of intracellular calcium concentration. There is some evidence that such biochemical second messengers (including cAMP, inositol trisphosphate, and perhaps diacylglycerol (DG) and cGMP) can participate in the control of membrane ion channels - that are not directly gated by receptors (see below). The metabolic effects of these substances are undoubtedly very diverse, however, with their signals being amplified through regulation of intracellular protein kinases, which can alter the properties of a number of intracellular proteins. At least some of their roles appear to apply to signal transduction in cells that are not overtly electrically excitable, as well as to that in neurones. Both amines and peptide hormones can modify the synthesis of secretory products in receptive cells. Such actions, possibly at the level of metabolic processing, or more likely at the level of the genome, may be mediated by second messengers or intracellular hormone-receptor complexes. Complexes of peptide hormone with receptor can be rapidly internalised from cell surface membranes [25] and may carry an intracellular signal. Receptors for some hormones (such as somatostatin) have surprisingly been found at high concentrations in the cytosolic fraction, not just on cell surface membranes [56]. Elevated intracellular calcium concentrations can be a sufficient signal to stimulate messenger RNA synthesis (for example prolactin mRNA in GH_3 clonal pituitary cells [79]). Raised intracellular cAMP levels can also stimulate transcription of genes for prolactin [43] and growth hormone [5], perhaps by phosphorylation of nuclear histone-associated proteins.

There appear, therefore, to be many levels on which neurotransmitters are likely to modify the activity of receptive cells, acting over a range of time scales. The core principle of all these phenomena is the interaction of the chemical messenger with a specific receptor site (usually on the cell surface) and a coupling of this event to message transduction in order to yield a significant biochemical or electrical signal.

Approaches to the Study of Central Neurotransmitter Action

Investigation of the mechanism of action of a central neurotransmitter requires a simplified system in which to examine quantifiable responses to a stimulus such as administration of the putative transmitter in question. Both electrophysiological and biochemical techniques have made important contributions, although these must also employ a pharmacological approach (such as the use of specific agonists and antagonists) to define the receptors involved. Intracellular recordings of neurone membrane potential can reveal the polarity of conductance changes, and the reversal potential of the response may be taken to infer the ionic species primarily mediating the response. Voltage-clamp techniques can further define the kinetic properties of particular ionic channels that might be involved in responses. The use of in vitro brain slice preparations and neuronal cultures (see for example [8, 61] has made it easier to perform quantitative pharmacology and manipulate channel function by ion substitutions or drugs.

The advent of two new biophysical techniques, fluctuation analysis and patch clamp (see [38] for review) has made it possible to investigate transmitter effects on the behaviour of individual ion channels. The mean channel conductance, open time and opening frequency can be derived. Different agonists appear to open channels for characteristically different times [7]. A number of centrally active drugs have further been shown by this approach to modulate either the open time or the opening frequency of channels activated by particular neurotransmitters [70]. Single ion-channel events in small regions of cell membrane can be observed in response to transmitters by patch-clamp techniques. Examples include actions of ACh on Na^+ channels at the neuromuscular junction [45], of 5-hydroxytryptamine (5-HT) on K^+ channels in sensory neurones of the invertebrate *Aplysia* [63], and β-noradrenergic effects on Ca^{2+} channels in the heart [55].

There are also a great variety of biochemical approaches available for studying neurotransmitter action. One of the most basic requirements in studying the action of a new putative transmitter is to establish the presence of specific receptor sites. Many laboratories have used the technique of radioligand binding to study the properties of specific high-affinity binding sites for radiolabelled transmitters or relevant drugs. This information is severely restricted, however, as it describes only the affinity of binding to a recognition site and nothing of the ensuing response mechanism. Occasionally, much more subtle information can be gained from binding studies, where some receptor complex contains a number of different recognition sites, capable of influencing each others' properties. The (A-type) receptor complex for γ-aminobutyric acid (GABA) contains sites that recognise GABA, benzodiazepine and certain convulsants (Fig. 1). Allosteric influences of one of these ligands on the binding of another provide a powerful approach to understanding the molecular interactions that occur within the complex [11, 37].

Aspects of response mechanisms can be studied more directly by other biochemical means. The production of intracellular putative second messengers, such as cAMP, in response to neurotransmitters is widely studied. Since the discovery of cAMP in the late 1950s, experiments measuring changes in its concentration or changes in the rate of activity of its synthetic enzyme, adenylate cyclase, have extensively characterised the properties of this transduction system [14, 58]. With the help of bacterial toxins, regulatory proteins that couple neurotransmitter receptors to adenylate cyclase have been discovered and character-

Fig. 1. The GABA$_A$/benzodiazepine/barbiturate receptor complex. The purified protein (mol. wt. in the order of 55 K) carries recognition sites for GABA and a number of pharmacological agents in three main domains. Binding to these sites exerts an allosteric influence on properties of the other sites [24]: for example, benzodiazepines and barbiturates enhance GABA binding, GABA and barbiturates enhance benzodiazepine binding, and the displacement of TBPS binding by GABA is facilitated by benzodiazepines [25]. There is evidence suggesting that the native receptor may be present as a cooperatively interacting dimer or tetramer. There are also suggestions of heterogeneous classes of GABA$_A$ receptor complex [25, 45]. *Wavy lines*, signals; **DBI** (an endogenous protein: "Diazepam Binding Inhibitor"); **TBPS** (a non-competitive, channel-blocking GABA$_A$ antagonist: t-butyl-bicyclophosphorothionate)

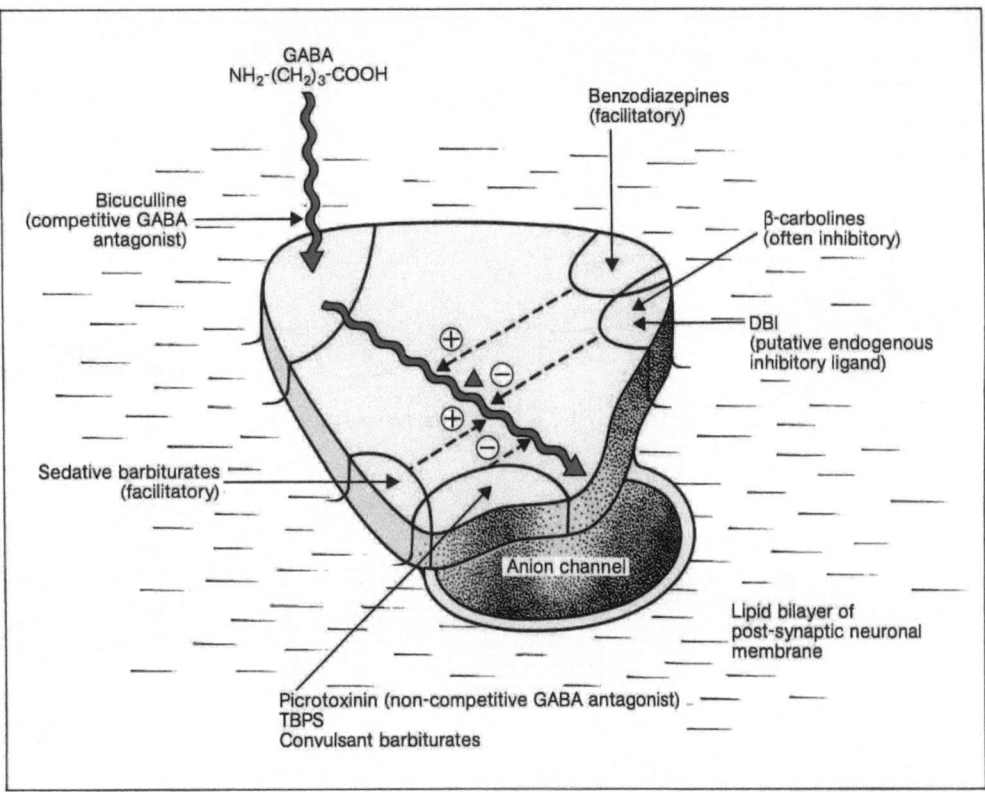

Fig. 2. Adenylate cyclase response mechanisms. Both stimulatory and inhibitory classes of receptors can modulate adenylate cyclase activity. Receptors are coupled to the enzyme through coupling GTPase proteins (N_s and N_i respectively). The bacterial toxins cholera toxin and pertussis toxin cause persistent activation of the cyclase by ADP-ribosylating sites on N_s and N_i, preventing dissociation of GDP and association of GTP respectively. The diterpene forskolin activates the enzyme directly and cAMP causes activation of previously inactive protein kinase A. This will phosphorylate a spectrum of intracellular proteins, apparently including a number of ion channels. *Wavy lines*, signals; GTP: guanosine 5'-triphosphate; GDP: guanosine 5'-diphosphate; ATP: adenosine 5'-triphosphate; cAMP: adenosine 3',5'-cyclic monophosphate; yellow boxes, pharmacological tools; **R**: regulatory and **C**: catalytic subunits of protein kinase A

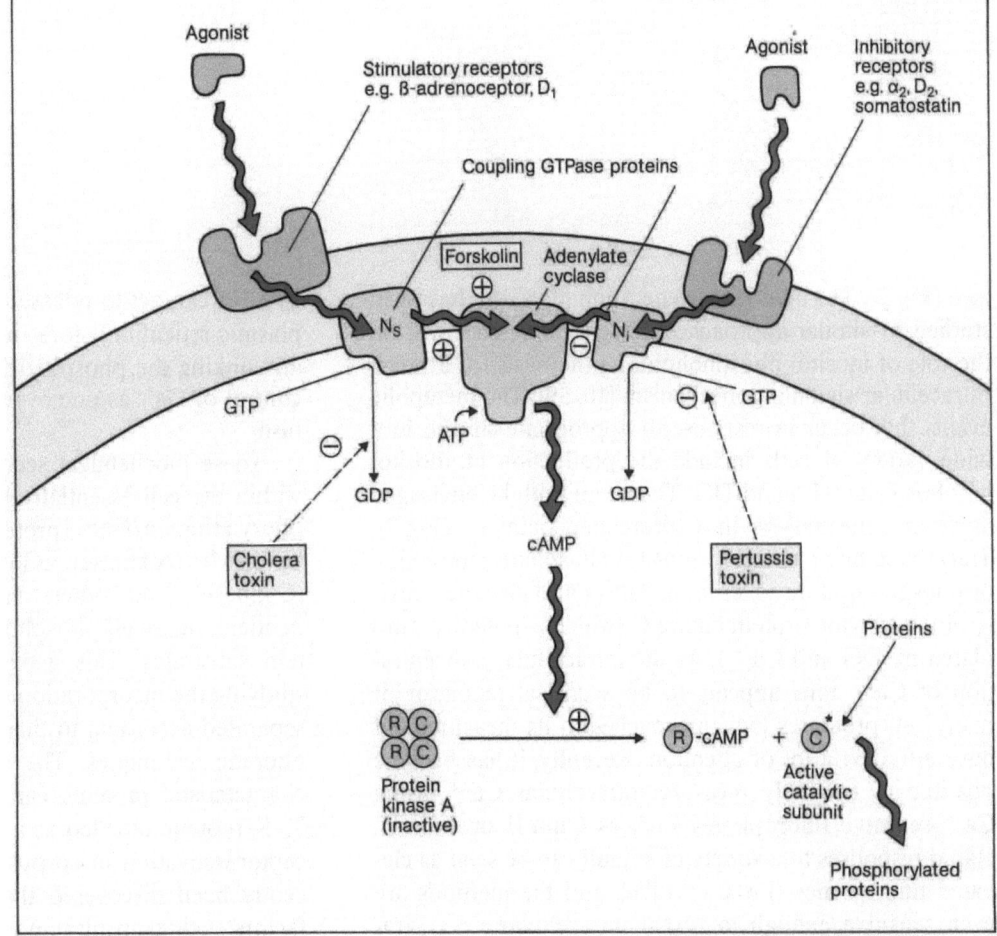

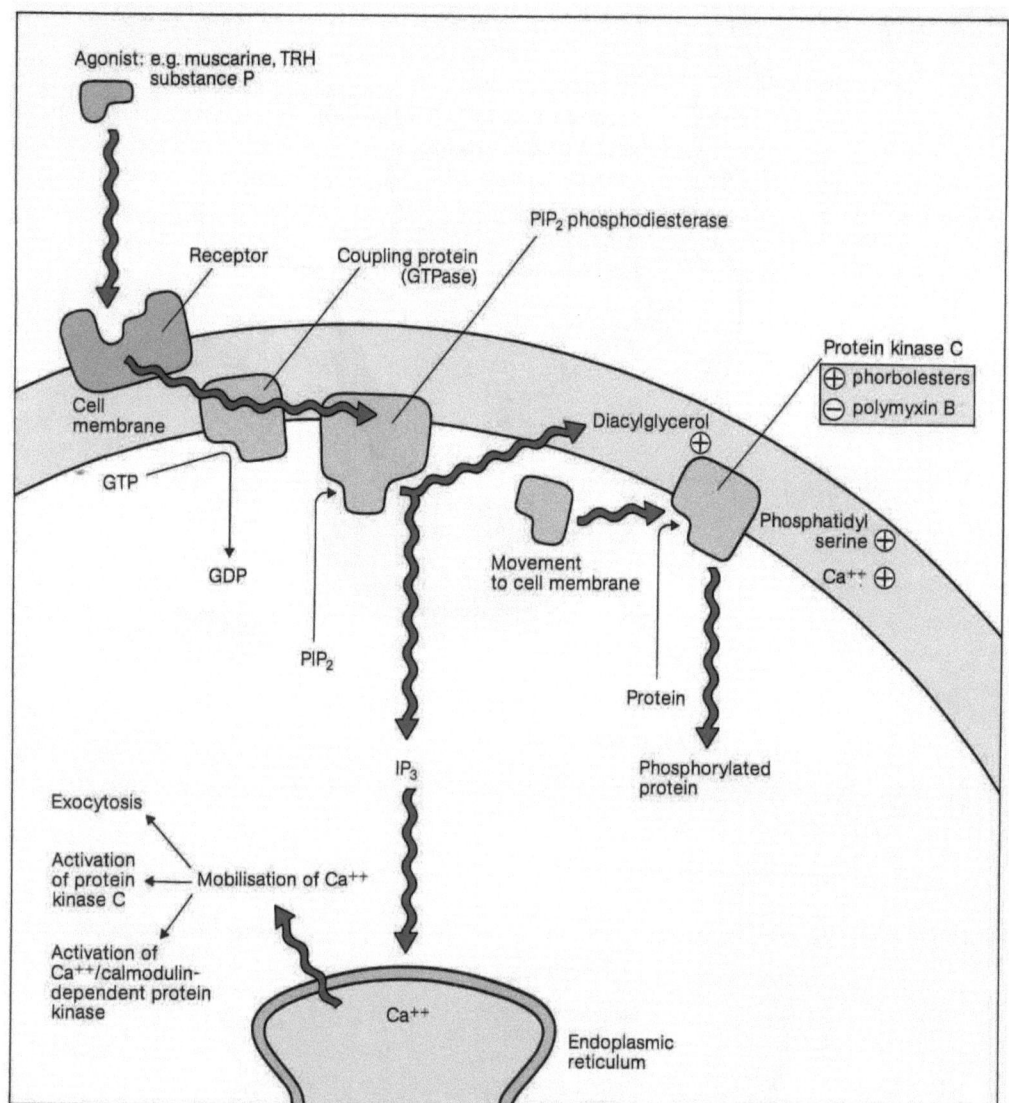

Fig. 3. Phosphoinositide/protein kinase C response mechanism. Receptor occupancy by agonist triggers the G protein (a coupling GTPase) to activate phosphodiesterase. This cleaves phosphatidylinositol 4,5-bisphosphate (PIP$_2$) to inositol triphosphate (IP$_3$) and diacylglycerol (DG), which act as potentially independent arms of the dual response mechanism. The 1,4,5 isomer of IP$_3$ mobilises Ca^{2+} from intracellular stores, which will cause exocytosis and phosphorylation of proteins by kinase C and calmodulin-dependent kinase. There may be additional functions of this and other inositol phosphates. DG activates protein kinase C, resulting in phosphorylation of appropriate proteins, perhaps including membrane ion channels. *Wavy lines*, signals; Abbreviations as in figure 2; yellow box, pharmacological tools

ised (Fig. 2). The analogous generation of cGMP has been studied by similar approaches. Recent interest centres on the role of inositol phospholipids as the basis for a novel intracellular signalling mechanism [10, 39]. The metabolic events that occur in response to appropriate stimuli in a wide variety of cells include the production of inositol trisphosphate (IP$_3$) and DG. These can both be envisaged as second messengers in a bifurcating pathway (Fig. 3). Tracer labelling can follow inositol phosphate production or phospholipid turnover in responses and also the activity of the enzyme protein kinase C (which is potently stimulated by DG and Ca^{2+}). As the intracellular concentration of Ca^{2+} ions appears to be a critical regulator of many cell processes [54], approaches to its measurement have attracted a lot of attention. Recently, it has become possible to routinely measure intracellular Ca^{2+} using Ca^{2+}-selective fluorophores such as Quin II or Fura II. Rapid responses to a variety of stimuli can be seen as elevated fluorescence (i.e. Ca^{2+}) [76], and the methods are even sensitive enough to reveal intracellular Ca^{2+} gradients [80]. A decisive series of experiments [68] revealed that IP$_3$ can act to release Ca^{2+} from intracellular (endoplasmic reticulum) stores into the cytosol, thus conceptually linking the phosphatidylinositol response cascade to control of Ca^{2+} as one overall signal transduction mechanism.

These biochemical second messengers appear to act within the cell essentially by activation of protein phosphorylating enzymes (protein kinases). Variants activated by cAMP (A kinase), cGMP (G kinase), DG and Ca^{2+} (C kinase), and others such as Ca^{2+}/calmodulin-dependent kinase all phosphorylate different spectra of protein substrates. This activity can be well followed by studying the incorporation of ^{32}P into proteins that can be separated according to their molecular weight by electrophoretic techniques. The phosphorylation of particular characteristic proteins can be observed (for example a 32-K protein, labelled as a consequence of dopamine receptor activation in corpus striatum [78]). It has very recently been discovered that receptors for some growth factors such as insulin and epidermal growth factor actually themselves have kinase activity. Upon activation, these receptors phosphorylate a number of intracellular substrates (including themselves) [62]. Whether such a mechanism could operate for central neurotransmitters, too, is not clear, but insulin receptors, for example, have been clearly identified within the brain [24].

A variety of other techniques can provide useful information on mechanisms of transmitter action. Fluorescent potential-sensitive dyes and radiolabelled lipophilic cations which distribute across cell membranes according to their potential form the basis for new approaches. The movement of particular ions can be easily followed by radiolabelled ion flux experiments. In many cells, transmitter-related events result in marked alteration of the output of secretory products from that cell. We have extensively employed such models in both CNS and endocrine tissue to study the mechanism of action of GABA and other receptors [4, 41, 42].

Mechanisms of Neurotransmitter Action

Direct Gating of Ion Channels

The effect of ACh at the neuromuscular junction is a rapid depolarisation of postsynaptic membrane by a rapid influx of sodium ions. Certain snake toxins recognise elements of the receptor/ion-channel complex and have been employed both pharmacologically and as ligands to assist in purification of the binding site(s).

This 'nicotinic' cholinergic receptor comprises several distinct subunit proteins - α, β, γ, δ - which are assembled in pentagonal symmetry, forming in the centre what appears to be a protected polar channel for the passage of sodium ions [67] (Fig. 4). The genetic sequences of each subunit have been determined by DNA cloning techniques. This has revealed the protein sequences and allowed computer prediction of the ways in which they are likely to fold through the membrane [48]. The pentagonal subunit scheme has been confirmed by direct observation [12], and the minimum reassembly of subunits necessary to reconstitute functional channels is being studied [60].

A-type receptors for the central transmitter GABA are also considered to directly control an ionic conductance, this time for chloride ions. At many sites this allows a passive influx of chloride, hyperpolarising the cell, but chloride is distributed quite close to equilibrium across the membrane, and some cells (or their processes) may show a reversed (outward) chloride gradient, where GABA can act in a depolarising fashion. In some cases both responses can be detected in the same cells, pointing to two distinct classes of $GABA_A$ receptor with somewhat different ionic mechanisms [3, 40]. The free movement of chloride ions, collapsing membrane resistance, will also diminish excitability in response to other inputs. A number of groups have purified proteins which carry recognition sites for GABA, benzodiazepine and convulsants like picrotoxin, all on apparently the same molecule [50, 65]. Attempts to clone its sequence are in progress. Although convulsants like picrotoxin are considered to influence chloride flux in $GABA_A$ responses rather than to interact with GABA recognition sites as such, it cannot be said yet that the entire receptor-channel complex resides within this one protein. This requires attempts to reconstitute functional receptor complexes from isolated proteins. Glycine and some excitatory amino acids may also directly gate ion conductances, but the majority of transmitter receptors appear to have an indirect action.

Indirect Gating of Ion Channels: a Role for Chemical Second Messengers

Many neurotransmitters can modify the properties of ion channels involved in action potential generation or recovery by what appear to be indirect means. The extended time course of a range of slow postsynaptic potentials has been suggested as being consistent with the involvement of a chemical second messenger, rather than direct channel gating [23]. It appears that many different types of voltage-sensitive channels can be influenced by different neurotransmitters. The opening of action potential Ca^{2+} channels in cardiac muscle is facilitated by β-adrenoreceptor agonists [75]. Since β-adrenergic drugs are known to stimulate cAMP production, it was hypothesised that cAMP was an intermediate in the modulation of Ca^{2+} conductance. This has been substantially confirmed by pharmacological manipulations of cAMP levels and by direct injection of cAMP into cells [64]. As mentioned above, cAMP is considered to act in general by stimulation of protein kinase A. The idea was proposed by Greengard [22] that cyclic nucleotides could act as mediators of slow synaptic potentials by inducing protein kinase to phosphorylate membrane ion channels. Firm support for such a mechanism has been provided by experiments in which the Ca^{2+} conductance changes produced by β-agonists could be mimicked by microinjection of the purified catalytic subunit of protein kinase A [52]. There is also good evidence that cAMP mediates the production by 5-HT of slow excitatory postsynaptic potentials in sensory neurones of Aplysia [64]. Electrophysiological studies have indicated that the inhibition of a tonically active outward K^+ current (I_s) is responsible. 5-Hydroxytryptamine increases cAMP levels in these neurones, and pharmacological elevation of intracellular cAMP or injection of protein kinase A catalytic subunit can reproduce the K^+ channel blockade by 5-HT. Microinjection of a specific inhibitor of protein kinase A can prevent the action of 5-HT. The involvement of a diffusable second messenger is also implicated by the appearance of conductance changes within the zone of patch-clamped membrane when 5-HT or cAMP analogues are applied to other parts of the cell. In Aplysia neurone R15, a different K^+ current, the "inward rectifier", is reported to be facilitated by 5-HT, also via a cAMP-dependent

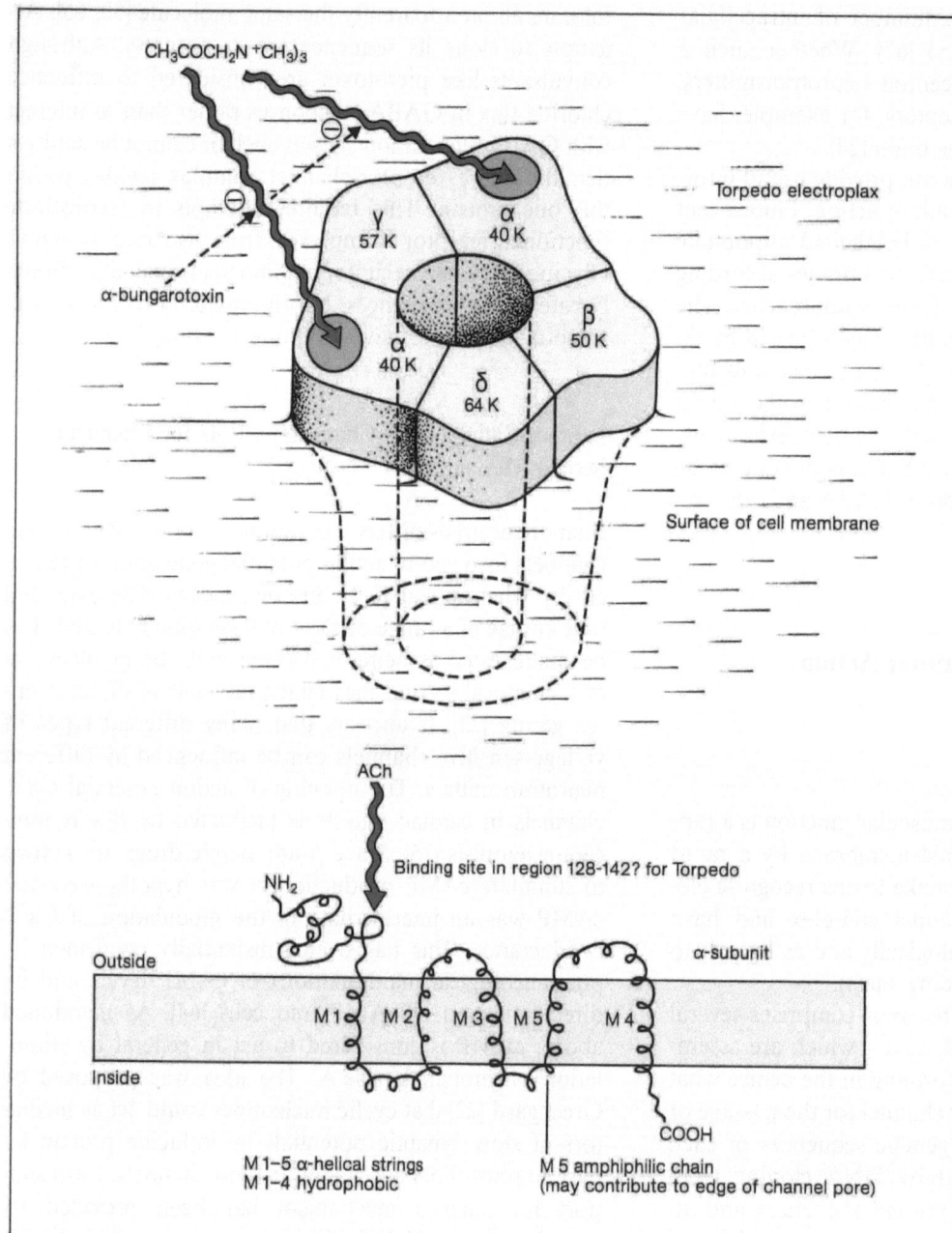

Fig. 4. Structure of nicotinic cholinergic receptor and its ion channel. The primary sequence of all subunits ($\alpha_2 \beta \gamma \delta$) for **Torpedo** electric organ has been derived by DNA cloning techniques. There is considerable homology at the amino acid level although each subunit is coded by separate genes. The hydrophobicity profiles (and predicted membrane-spanning regions) of each subunit are very similar, and each probably contributes an amphiphilic chain (M 5) to the side of the ion channel. Acetylcholine and the snake venom α-bungarotoxin bind the α subunit. Toxin affinity columns have been critical in purification of the receptor. The quaternary structure of the receptor shows exquisite pentagonal symmetry. **Wavy lines**, signals

mechanism [18]. In rat hippocampal pyramidal cells, activation of β-adrenoreceptors inhibits the accommodation of the firing rate normally occurring with a train of successive stimuli [35]. β-Agonists and cAMP appear to act by blocking a Ca^{2+}-activated K^+ conductance by a mechanism other than that of restricting Ca^{2+} entry. Indeed, the catalytic subunit of kinase A can facilitate opening of a Ca^{2+}-dependent K^+ channel in membrane patches of *Helix* neurones [20]. Yet another type of K^+ current (the transient outward current, I_A) is powerfully inhibited by cAMP analogues in *Aplysia* bag cells [69], although whether this participates in the action of a neurotransmitter is not clear. What is clear from this array of channel modifications is that cAMP-stimulated protein kinase A is potentially capable of many diverse actions on ion-channel properties. Indeed, large numbers of intracellular proteins are phosphorylated in response to 5-HT in *Aplysia* R15 neurones for example [33]. It seems likely that other factors in the cell, perhaps the activation of several second messenger signals by transmitters, will regulate which changes in ion-channel properties are allowed to predominate.

There is much less evidence for cGMP acting as a second messenger in the indirect gating of ion channels. This has, however, been suggested for muscarinic cholinergic inhibition of Ca^{2+} conductance in cardiac cells [73] and for the light-induced closure of Na^+ channels in retinal rods [36], and substrate phosphorylation by cGMP-dependent protein kinase has been characterised in cerebellar Purkinje cells [16].

Many neurotransmitter actions appear not to involve cyclic nucleotides, but seem instead to centre on the regulation of intracellular Ca^{2+} concentration [54]. In many of these cases the metabolism of membrane inositol phospholipids is rapidly stimulated. The clear correlation between those receptors stimulating phosphoinositide turnover and those causing elevation of intracellular Ca^{2+} led to the hypothesis that this metabolic change is the critical regulator involved [39]. In various tissues acetylcholine, 5-HT, vasopressin, angiotensin, thyrotropin-releasing hormone (TRH), substance P, bradykinin, bombesin and quisqualate have been shown to evoke hydrolysis of phosphatidylinositol bisphosphate (PIP_2), which is now considered to be the critical substrate for the signalling mechanism [10]. The resulting signal appears to be a dual one. Both IP_3 and DG are produced. The former can act to transiently mobilise intracellular Ca^{2+} stores, and this raised Ca^{2+} level may act as a critical signal for secretory responses, perhaps by means of the two kinases (Ca^{2+}/calmodulin-dependent kinase and protein kinase C) that it activates. Indeed, IP_3 has been shown in permeabilised human platelets to increase phosphorylation of a number of proteins that are substrates for each of these kinases [32]. Diacylglycerol on the other hand activates primarily protein kinase C. In some cells, this itself is an adequate second messenger to elicit responses, with no requirement for a raised intracellular Ca^{2+} level [57]. Tumour-promoting phorbol esters act relatively directly on protein kinase C to activate it, whilst a few substances such as polymyxin B act as selective antagonists [47]. In a number of cell types, phorbol esters potentiate responses evoked by elevating intracellular Ca^{2+} levels, suggesting synergy between the two arms of the second messenger cascade [47]. In some cases, for example the activation of T-lymphocytes by antigens, the response can be mimicked only by a combination of Ca^{2+} elevation and protein kinase C activation and by neither alone [74]. Membrane calcium currents in *Aplysia* neurones can be facilitated by phorbol esters [15]. Protein kinase C can act to inhibit responses, too. The secretory response of gonadotropes to LHRH is elevated by polymyxin B [28]. Phorbol esters (at slightly higher concentrations than those that produce synergy) will inhibit responses to antigen or elevated Ca^{2+} in basophilic leukaemia (2H3) cells. In this case, the antigen-induced elevation of Ca^{2+} concentration is abolished by phorbol esters [59]. There is some evidence that such inhibitory actions may appear at high agonist concentrations (at least for TRH action on GH_3 cells) and that the kinase-C arm of the pathway may therefore provide a negative feedback role in regulating Ca^{2+}-mobilisation responses [17]. It may be that kinase C can modulate Ca^{2+} mobilisation from intracellular stores by IP_3 or Ca^{2+} influx through membrane ion channels. Phosphorylation of transmitter-synthesising enzymes and certain membrane receptors and transporters has indeed been demonstrated in response to phorbol esters [2, 30, 72, 82]. Interactions with other second messenger systems can also be seen, such as a facilitation by phorbol esters of β-adrenoreceptor-induced cAMP synthesis [71]. Evidence for the direct involvement of the phosphoinositide pathway in electrical responses of excitable cells is much more tenuous, however. Nevertheless, intracellular injection of IP_3 into *Xenopus* oocytes was shown to mimic the depolarising calcium-dependent chloride current evoked by muscarinic cholinergic agonists [51]. The MI class of (pirenzipine-sensitive) muscarinic receptors appear to be coupled to the phosphoinositide response pathway [21], and in rat superior cervical ganglion they seem to mediate the muscarinic slow depolarisation [46]. This involves inhibition of a delayed outward K^+ current which is not the classical delayed rectifier, sensitive to tetraethylammonium (TEA). In frog sympathetic ganglia, and to a lesser extent in mammalian ganglia and hippocampus, the current involved has been characterised. It appears to be a novel class of K^+ current, designated the 'M current' [13]. The same current is inhibited by an LHRH-like peptide in frog sympathetic ganglia, but this may not account fully for the peptide-induced depolarisation [27]. Although LHRH can evoke some signs of a phosphoinositide response in mammalian gonadotropes [44], we can find no evidence that M-current blockade mediates the action of LHRH on LH secretion [42]. Substance P, which evokes a marked phosphoinositide response in the parotid gland [9], appears to block a K^+ current with many similar properties to the M current in mouse spinal cord neurones [49]. It appears, however, that there is not just a single class of current gated by phosphoinositides. In other neurones, substance P appears not to modulate M currents, but to block the Rb^+-sensitive, inwardly rectifying K^+ channel [66]. Different laboratories have implicated blockade of either the transient outward current or a TEA-sensitive current in TRH action [6, 29]. Suppression of this transient outward current may also participate in the action of the α-adrenoreceptor (another phosphoinositide-linked receptor) [1]. It may well be that a number of channels and other intracellular regulator sites can be substrates for the cascade of phosphorylation induced in the phosphoinositide response.

In very few cases, therefore, is it completely clear in what way neurotransmitters induce electrical (or metabolic) signals in receptive cells. Certainly, in some cases, the direct gating of membrane ion channels seems to be involved. More commonly, however, (particularly with neuropeptides) it seems to be the indirect gating of voltage-dependent channels that is critical. The chemical second messengers that appear commonly to be involved permit a diverse control of many aspects of cellular function from one initial signal. Changes in membrane excitability, specific modulation of responses to other inputs, and broad-ranging metabolic changes appear to be possible. These cellular transduction mechanisms that have so far been identified, together with the factors regulating interplay between them, are the subject of intensive research.

Many neurones are now well documented as containing more than one neurotransmitter [34]. If each is releasable and then detectable by a given cell, a further level of complexity may be added to the mode of information transfer. One substance may modify responsiveness to its co-transmitter. Clearly, therefore, chemical neurotransmission permits the encoding of far more subtle information than could be achieved with digital processing of similar anatomical complexity.

References

1. Aghajanian GK (1985) Modulation of a transient outward current in serotonergic neurones by α_1-adrenoceptors. Nature 315: 501–503
2. Albert KA, Helmer-Matyjek E, Nairn AC, Muller TH, Haycock JW, Greene LA, Goldstein M, Greengard P (1984) Calcium/phospholipid-dependent protein kinase (protein kinase C) phosphorylates and activates tyrosine hydroxylase. Proc Natl Acad Sci USA 81: 7713–7717
3. Alger BE, Nicoll RA (1982) Pharmacological evidence for two kinds of GABA receptor on rat hippocampal pyramidal cells studied in vitro. J Physiol (Lond) 328: 125–141
4. Anderson RA, Mitchell R (1986) Biphasic effect of $GABA_A$ receptor agonists on prolactin secretion: evidence for two types of $GABA_A$ receptor complex on lactotrophes. Eur J Pharmacol 124: 1–9
5. Barinaga M, Bilezikjian LM, Vale WW, Rosenfeld MG, Evans RM (1985) Independent effects of growth hormone releasing factor on growth hormone release and gene transcription. Nature 314: 279–281
6. Barker JL, Dufy B, Owen D, Segal M (1983) Excitable membrane properties of cultured CNS neurons and clonal pituitary cells. Cold Spring Harbor Symp Quant Biol 48: 259–268
7. Barker JL, Mathers DA (1981) GABA analogues activate channels of different duration on cultured mouse spinal neurons. Science 212: 358–361
8. Barker JL, Ransom BR (1978) Amino acid pharmacology of mammalian central neurones grown in tissue culture. J Physiol (Lond) 280: 331–354
9. Berridge MJ, Dawson RMC, Downes CP, Heslop JP, Irvine RF (1983) Changes in the levels of inositol phosphates after agonist-dependent hydrolysis of membrane phospholipids. Biochem J 212: 473–482
10. Berridge MJ, Irvine RF (1984) Inositol trisphosphate, a novel second messenger in cellular signal transduction. Nature 312: 315–321
11. Braestrup C (1982) Neurotransmitters and CNS disease: anxiety. Lancet ii: 1030–1034
12. Brisson A, Unwin PNT (1985) Quaternary structure of the acetylcholine receptor. Nature 315: 474–477
13. Brown DA (1983) Slow cholinergic excitation - a mechanism for increasing neuronal excitability. Trends Neurosci 6: 302–307
14. Codina J, Hildebrandt JD, Sekura RD, Birnbaumer M, Bryan J, Manclark R, Iyengar R, Birnbaumer L (1984) N_s and N_i, the stimulatory and inhibitory regulatory components of adenyl cyclases: purification of the human erythrocyte proteins without the use of activating regulatory ligands. J Biol Chem 259: 5871–5886
15. DeRiemer SA, Strong JA, Albert KA, Greengard P, Kaczmarek LK (1985) Enhancement of calcium current in Aplysia neurones by phorbolester and protein kinase C. Nature 313: 313–316
16. Detre JA, Nairn AC, Aswad DW, Greengard P (1984) Localization in mammalian brain of G-substrate, a specific substrate for guanosine 3',5'-cyclic monophosphate-dependent protein kinase. J Neurosci 4: 2843–2849
17. Drummond AH (1985) Bidirectional control of cytosolic free calcium by thyrotropin-releasing hormone in pituitary cells. Nature 315: 752–755
18. Drummond AH, Benson JA, Levitan IE (1980) Serotonin-induced hyperpolarisation of an identified Aplysia neuron is mediated by cyclic AMP. Proc Natl Acad Sci USA 77: 5013–5017
19. Dunlap K, Fischbach GD (1981) Neurotransmitters decrease the calcium conductance activated by depolarisation of embryonic chick sensory neurones. J Physiol (Lond) 317: 519–535
20. Ewald DA, Williams A, Levitan IB (1985) Modulation of single Ca^{2+}-dependent K^+ channel activity by protein phosphorylation. Nature 315: 503–506
21. Gonzales RA, Crews FT (1984) Characterisation of the cholinergic stimulation of phosphoinositide hydrolysis in rat brain slices. J Neurosci 4: 3120–3127
22. Greengard P (1978) Phosphorylated proteins as physiological effectors. Science 199: 146–152
23. Hartzell HC (1981) Mechanisms of slow synaptic potentials. Nature 291: 539–544
24. Havrankova J, Roth J, Brownstein M (1978) Insulin receptors are widely distributed in the central nervous system of the rat. Nature 272: 827–829
25. Hazum E, Cuatrecasas P, Marian J, Conn PM (1980) Receptor-mediated internalisation of fluorescent gonadotropin-releasing hormone by pituitary gonadotropes. Proc Natl Acad Sci USA 77: 6692–6695
26. Iversen LL (1983) Neuropeptides - what next? Trends Neurosci 6: 293–294
27. Jan YN, Jan LY (1983) An LHRH-like peptidergic neurotransmitter capable of action at a distance in autonomic ganglia. Trends Neurosci 6: 320–325
28. Johnson M, Mitchell R, Fink G (1986) The priming effect of LHRH: is protein kinase C involved? Proc Br Endocrine Soc, April
29. Kaczorowski GJ, Vandlen RL, Katz GM, Reuben JP (1983) Regulation of excitation-secretion coupling by thyrotropin-releasing hormone (TRH): evidence for TRH receptor-ion channel coupling in cultured pituitary cells. J Membr Biol 71: 109–118
30. Kelleher DJ, Pessin JE, Ruoho AE, Johnson GL (1984) Phorbolester induces desensitisation of adenylate cyclase and phosphorylation of the β-adrenergic receptor in turkey erythrocytes. Proc Natl Acad Sci USA 81: 4316–4320
31. Kelly JS (1982) Electrophysiology of peptides in the central nervous system. Br Med Bull 38: 283–290
32. Lapetina EG, Watson SP, Cuatrecasas P (1984) Myo-inositol 1,4,5-triphosphate stimulates protein phosphorylation in saponin-permeabilised human platelets. Proc Natl Acad Sci USA 81: 7431–7435
33. Levitan IB, Lemos JT, Novak-Hofer I (1983) Protein phosphorylation and the regulation of ion channels. Trends Neurosci 6: 496–499
34. Lundberg JM, Hökfelt T (1983) Coexistence of peptides and classical transmitters. Trends Neurosci 6: 325–333
35. Madison DV, Nicholl RA (1982) Noradrenaline blocks accommodation of pyramidal cell discharge in the hippocampus. Nature 299: 636–638
36. Matthews HR, Torre V, Lamb TD (1985) Effects on the photoresponse of calcium buffers and cyclic GMP incorporated into the cytoplasm of retinal rods. Nature 313: 582–584
37. McAllister-Williams RH, Mitchell R (1985) Benzodiazepines regulate coupling to anion channels in only some $GABA_A$ receptor complexes. Br J Pharmacol 84: 60P
38. McBurney RN (1983) New approaches to the study of rapid events underlying neurotransmitter action. Trends Neurosci 6: 297–302
39. Michell RH (1975) Inositol phospholipids and cell surface receptor function. Biochim Biophys Acta 415: 81–147
40. Mitchell R, Anderson RA (1985) Antagonism by strychnine differentiates two subtypes of $GABA_A$ receptor complex. Biochem Soc Trans 13: 1216–1217
41. Mitchell R, Anderson RA (1985) Does an anion channel mediate the action of K opioid receptors? Regul Pept [Suppl] 4: 191–196

42. Mitchell R, Ogier S-A, Johnson M, Cleland A, Bennie J, Fink G (1986) Evidence for sex differences in GnRH receptors and mechanism of action. In: Neuroendocrine molecular biology. Ed: Fink G, Harmar AJ & McKerns KW Plenum, London, pp 91-100
43. Murdoch GH, Rosenfeld MG, Evans RM (1982) Eukaryotic transcriptional regulation and chromatin-associated phosphorylation by cyclic AMP. Science 218: 1315-1317
44. Naor Z, Amsterdam A, Catt KJ (1984) Binding and activation of gondadotropin-releasing hormone receptors in pituitary gonadotropes. In: Hormone receptors in growth and reproduction. Ed: Fink G, Harmar AJ & McKerns KW Raven, New York, pp 113-124
45. Neher E, Sakmann B (1976) Single-channel currents recorded from membrane of denervated frog muscle fibres. Nature 260: 799-802
46. Newberry NR, Priestley T, Woodruff GN (1985) Pharmacological distinction between two muscarinic responses on the isolated superior cervical ganglion of the rat. Eur J Pharmacol 116: 191-192
47. Nishizuka Y (1984) The role of protein kinase C in cell surface signal transduction and tumour promotion. Nature 308: 693-698
48. Noda M, Takahashi H, Tanabe T, Toyosato M, Kikyotani S, Furutani Y, Hirose T, Takashima H, Inayama S, Miyata T, Numa S (1983) Structural homology of *Torpedo californica* acetylcholine receptor subunits. Nature 302: 528-532
49. Nowak LM, Macdonald RL (1982) Substance P: ionic basis for depolarising responses in cell culture. J Neurosci 2: 1119-1128
50. Olsen RW, Fischer JB, King RG, Ransom JY, Stauber GB (1984) Purification of the GABA/benzodiazepine/barbiturate receptor complex. Neuropharmacology 23 (7 B): 853-855
51. Oron Y, Dascal N, Nadler E, Lupu M (1985) Inositol 1,4,5-trisphosphate mimics muscarinic response in *Xenopus* oocytes. Nature 313: 141-143
52. Osterreider W, Brum G, Hescheler J, Trautwein W, Flockerzi V, Hofmann F (1982) Injection of subunits of cyclic AMP-dependent protein kinase into cardiac myocytes modulates Ca^{2+} current. Nature 298: 576-578
53. Paupardin-Tritsch D, Colombaioni L, Deterre P, Gerschenfeld HM (1985) Two different mechanisms of calcium spike modulation by dopamine. J Neurosci 5: 2522-2532
54. Rasmussen H, Barrett PQ (1984) Calcium messenger system - an integrated view. Physiol Rev 64: 938-984
55. Reuter H (1983) Calcium channel modulation by neurotransmitters, enzymes and drugs. Nature 301: 569-574
56. Reyl-Desmars F, Lewin MJM (1982) Evidence for an intracellular somatostatin receptor in pancreas: a comparative study with reference to gastric mucosa. Biochem Biophys Res Comm 109: 1324-1331
57. Rink TJ, Sanchez A, Hallam TJ (1983) Diacyl glycerol and phorbol ester stimulate secretion without raising cytoplasmic free calcium in human platelets. Nature 305: 317-319
58. Rodbell M (1980) The role of hormone receptors and GTP-regulatory proteins in membrane transduction. Nature 284: 17-22
59. Sagi-Eisenberg R, Lieman H, Pecht I (1985) Protein kinase C regulation of the receptor-coupled calcium signal in histamine-secreting rat basophilic leukaemia cells. Nature 313: 59-60
60. Sakmann B, Methfessel C, Mishina M, Takahashi T, Takai T, Kurasaki M, Fukuda K, Numa S (1985) Role of acetylcholine receptor subunits in gating of the channel. Nature 318: 538-543
61. Schwartzkroin PA (1975) Characteristics of CA I neurons recorded intracellularly in the hippocampal in vitro slice preparation. Brain Res 85: 423-426
62. Sefton BM, Hunter T (1984) Tyrosine protein kinases. Adv Cyclic Nucleotide Protein Phosphorylation Res 18: 195-217
63. Siegelbaum SA, Camardo JS, Kandel ER (1982) Serotonin and cyclic AMP close single K^+ channels in *Aplysia* sensory neurones. Nature 299: 413-417
64. Siegelbaum SA, Tsien RW (1983) Modulation of gated ion channels as a mode of transmitter action. Trends Neurosci 6: 307-313
65. Sigel E, Barnard EA (1984) A α-aminobutyric acid/benxodiazepine receptor complex from bovine cerebral cortex: improved purification with preservation of regulatory sites and their interactions. J Biol Chem 259: 7219-7223
66. Stanfield PR, Nakajima Y, Yamaguchi K (1985) Substance P raises neuronal membrane excitability by reducing inward rectification. Nature 315: 498-501
67. Stevens CF (1985) Acetylcholine receptors; fivefold symmetry and the ε subunit. Trends Neurosci 8: 335-336
68. Streb H, Irvine RF, Berridge MJ, Schulz I (1983) Release of Ca^{2+} from a non-mitochondrial intracellular store in pancreatic acinar cells by inositol-1,4,5-trisphosphate. Nature 306: 67-69
69. Strong JA (1984) Modulation of potassium current kinetics in bag cell neurones of *Aplysia* by an activator of adenylate cyclase. J Neurosci 4: 2772-2783
70. Study RE, Barker JL (1981) Diazepam and (-) pentobarbital: fluctuation analysis reveals different mechanisms for potentiation of α-aminobutyric acid responses in cultured central neurones. Proc Natl Acad Sci USA 78: 7180-7184
71. Sugden D, Vanecek J, Klein DC, Thomas TP, Anderson WB (1985) Activation of protein kinase C potentiates isoprenaline-induced cyclic AMP accumulation in rat pinealocytes. Nature 314: 359-361
72. Takayama S, White MF, Lauris V, Kahn CR (1984) Phorbol esters modulate insulin receptor phosphorylation and insulin action in cultured hepatoma cells. Proc Natl Acad Sci USA 81: 7797-7801
73. Trautwein W, Taniguchi J, Noma A (1982) The effects of intracellular cyclic nucleotide and calcium on the action potential and acetylcholine response of isolated cardiac cells. Pflügers Arch 392: 307-314
74. Truneh A, Albert F, Golstein P, Schmitt-Verhulst A-M (1985) Early steps of lymphocyte activation bypassed by synergy between calcium ionophores and phorbol ester. Nature 313: 318-320
75. Tsien RW (1977) Cyclic AMP and contractile activity in the heart. Adv Cyclic Nucleotide Res 8: 363-420
76. Tsien RY, Pozzan T, Rink TJ (1982) T-cell mitogens cause early changes in cytoplasmic free Ca^{2+} and membrane potential in lymphocytes. Nature 295: 68-71
77. Tsunoo A, Konishi S, Otsuka M (1982) Substance P as an excitatory transmitter of primary afferent neurons in guinea-pig sympathetic ganglia. Neuroscience 7: 2025-2037
78. Walaas SI, Ouimet CC, Hemmings HC, Greengard P (1985) Dopamine regulated protein phosphorylation systems in the basal ganglia. Neurosci Lett [Suppl] (1985): 5409
79. White BA, Bauerle LR, Bancroft FC (1981) Calcium specifically stimulates prolactin synthesis and messenger RNA sequences in GH_3 cells. J Biol Chem 256: 5942-5945
80. Williams DA, Fogarty KE, Tsien RY, Fay FS (1985) Calcium gradients in single smooth muscle cells revealed by the digital imaging microscope using Fura-II. Nature 318: 558-561
81. Williams JT, Egan TM, North RA (1982) Enkephalin opens potassium channels on mammalian central neurons. Nature 299: 74-77
82. Witters LA, Vater CA, Lienhard GE (1985) Phosphorylation of the glucose transporter in vitro and in vivo by protein kinase C. Nature 315: 777-778

Clinical Relevance

Janice E. Christie

MRC Brain Metabolism Unit, Royal Edinburgh Hospital, Morningside, Edinburgh, United Kingdom

This section considers the involvement of neurotransmitters in the major mental illnesses and in degenerative disorders of the central nervous system: Parkinson's disease, Huntington's disease and Alzheimer-type dementia.

The Functional Psychoses

Abnormalities in central neurotransmitter systems have long been postulated as a cause for the functional psychoses: manic-depressive psychosis and schizophrenia. This idea was based firstly on the reversibility of psychosis and secondly on the actions of drugs both in inducing symptoms of psychosis and in the treatment of psychosis.

Let us consider first of all the reversible nature of the functional psychosis. Kraepelin made the fundamental division of the functional psychoses into manic-depressive psychosis, primarily a disturbance of mood in which there is eventual complete recovery after each episode of illness, and dementia praecox (schizophrenia), in which acute psychotic symptoms may remit but there is a gradual decline in the patient's functioning, with the development of a defect state characterised by blunted emotional responses, social withdrawal and loss of drive and initiative. The division of the functional psychoses into manic-depressive psychosis and schizophrenia is, however, oversimplified. There appears to be a group of patients who show both a disturbance of mood (affective symptoms) and some of the acute symptoms of schizophrenia. These schizoaffective patients tend to make a good recovery, and family studies suggest that their disturbance is more closely related to manic-depressive psychosis than to schizophrenia.

Schizophrenia

The symptoms of acute schizophrenia are hallucinations, which are normally auditory, and delusions (false beliefs which are held with conviction despite rational evidence to the contrary), the most common delusional beliefs having a paranoid content. There is also a disturbance of thinking with a loosening of associations; the links between phrases become so tenuous that meaning may be lost (thought disorder). The acute reversible phenomena of schizophrenia are often called "positive" symptoms and respond well to treatment with antipsychotic drugs, while the negative symptoms of the dementia-like defect state do not tend to respond to these drugs [11]. An abnormality in the function of a neurotransmitter system may explain positive symptoms, but it is likely that structural neuronal change underlies the permanent defect state of chronic schizophrenia [4]. This concept is supported by computerised tomography (CT) scan evidence of shrinkage of the brain with enlargement of the lateral ventricles in chronic schizophrenic patients [20]. The CT scan is a technique which uses a computer to combine X-ray readings taken from many different angles to represent cross sections of the brain. Furthermore, recent studies suggest that there are neuropathological changes in the temporal lobe, particularly in the entorhinal cortex and hippocampus.

The neurotransmitter which has been most closely linked to the acute symptoms of schizophrenia is dopamine. The "dopamine hypothesis" of schizophrenia suggests that the symptoms of schizophrenia are associated with either an overactivity of dopamine neurones or an increase in dopamine receptors. This hypothesis was based on the findings that (a) amphetamine intoxication could cause a psychosis indistinguishable from acute schizophrenia, and (b) all antipsychotic drugs (neuroleptics) have a common action - blockade of dopamine receptors.

Amphetamine interacts in a number of ways with dopamine neurones, increasing the release of dopamine and also potentiating the action of dopamine by inhibiting its re-uptake into the neurone. There are a large number of drugs, such as the phenothiazines, thioxanthenes and butyrophenones, which are effective in treating psychotic symptoms, and although they have many different actions, all are potent dopamine receptor antagonists. Furthermore, there is a high correlation between the mean clinical dose of neuroleptic drug required to treat schizophrenia and the efficacy in blocking dopamine receptors. At least two types of dopamine receptor exist: D_1 receptors, which are linked to stimulation of adenylate cyclase,

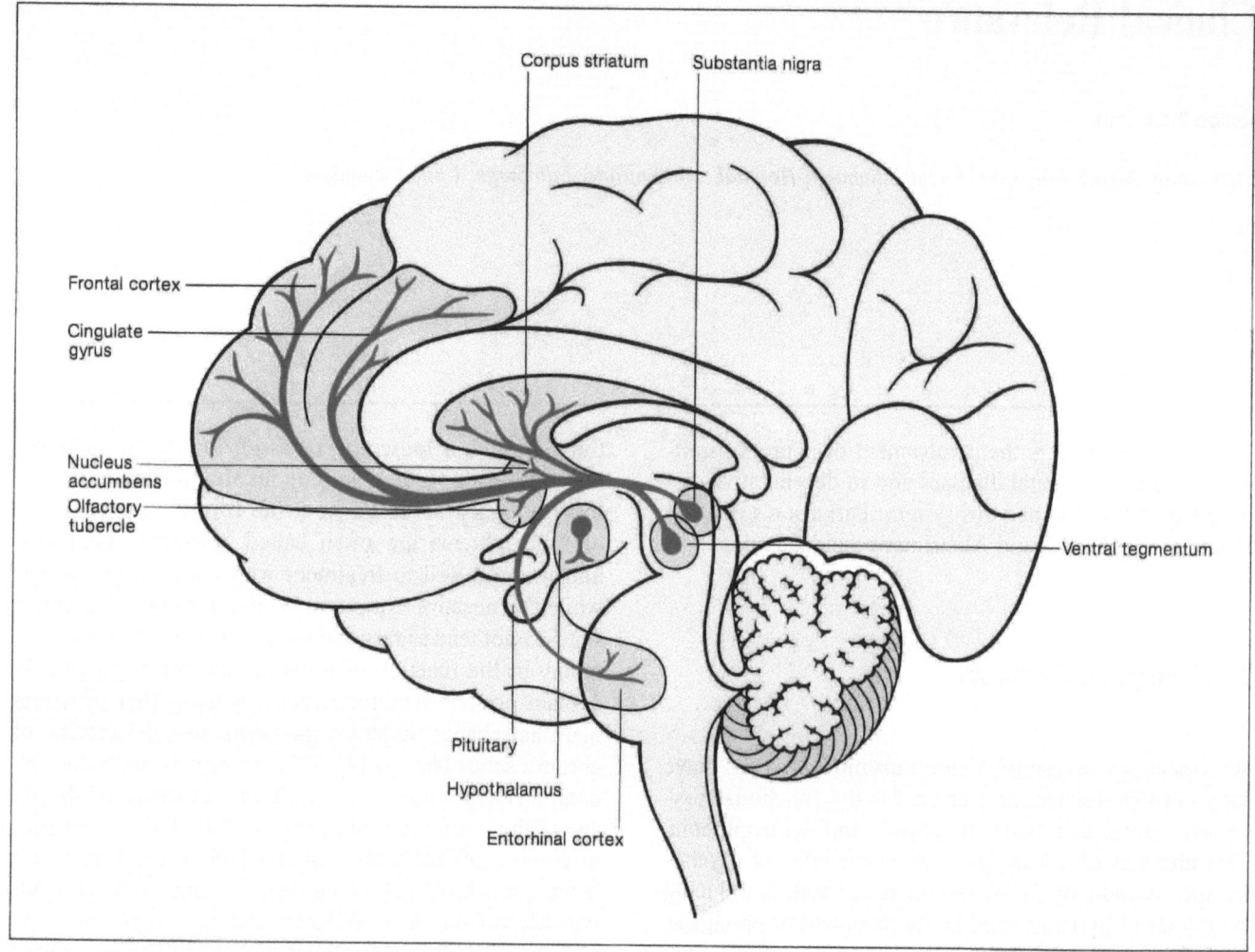

Fig. 1. The major dopamine pathways in the human brain. The dopamine cell bodies in the substantia nigra project to the corpus striatum, and a deficiency of dopamine in the striatum is associated with Parkinson's disease. The ventral tegmental dopamine cell bodies form the mesocortical and mesolimbic dopamine pathways, which project respectively to the frontal cortex and to a variety of limbic forebrain structures including the nucleus accumbens, the olfactory tubercle, the amygdala and the entorhinal cortex. The short dopamine pathway in the hypothalamus regulates hormone release from the pituitary gland; most importantly, dopamine inhibits prolactin release

and D_2 receptors, which are not associated with, or are inhibitory for adenylate cyclase. The antipsychotic effects are related to D_2 receptor blockade. It is important to note that these antipsychotic drugs are effective in treating psychotic symptoms irrespective of their cause and are not simply "antischizophrenic" drugs. They are used in the treatment of mania and the organic psychoses.

The 'dopamine hypothesis' of schizophrenia has been investigated in post mortem neurochemical studies. Unfortunately, interpretation of the results is problematic, because almost all schizophrenic patients have been treated with antipsychotic drugs which increase the number of dopamine receptors and it is difficult to separate drug effects from the effects of the illness. The finding of increased dopamine levels and increased D_2 receptor densities in schizophrenic brains may simply reflect drug treatment [12]. However, a recent study has reported a higher dopamine content in the left than in the right amygdala (a deep structure of the temporal lobe), and this result favours an illness interpretation because a drug effect would be expected to be symmetrical [13]. The mesocortical and mesolimbic dopamine systems which project to the frontal cortex and to limbic structures such as the nucleus accumbens, amygdala, olfactory tubercle and entorhinal cortex are more likely to be involved in the pathogenesis of schizophrenia than the nigrostriatal system, whose function is more clearly related to the initiation of movement and the degeneration of which is associated with Parkinson's disease (Fig. 1).

Manic-Depressive Psychosis

Manic-depressive psychosis is a disturbance of mood. The manic phase is characterised by elation, excitement, overactivity and grandiose ideas or delusions. The depressive phase is much more common, and patients may have recurrent depressive illnesses without ever being manic, so-called unipolar depression, whereas those who have both manic and depressive episodes are known as bipolar. Mania responds rapidly to treatment with antipsychotic drugs (dopamine receptor antagonists) and it is therefore suspected, as in schizophrenia, that there is overactivity of dopamine systems. In contrast, the neurotransmitters most closely linked to potential neurochemical causes of depression are noradrenaline, 5-hydroxytryptamine and possibly also acetylcholine.

Depression of mood in response to adverse circumstances is a universal phenomenon, but the diagnosis of a depressive illness is dependent on the presence of a cluster of physical symptoms which accompany the mood change. There is frequently sleep disturbance, characterised by awakening in the early hours of the morning and being unable to get back to sleep again. Appetite is poor and weight is lost. The depression is at its worst on first awakening in the morning and may gradually improve as the day goes on (diurnal variation of mood). There is loss of interest and poor concentration, and in more severe depressions retardation (slowing of thoughts and actions) or agitation occurs. These symptoms are characteristic of endogenous depression. A small number of severely depressed patients have psychotic symptoms, especially delusional beliefs, for example, that they are worthless, that they have sinned greatly and deserve to be punished or to die. Auditory hallucinations reflect these beliefs and the patients may hear a "voice" telling them that they are dirty or evil. Patients with many of the symptoms of endogenous depression tend to respond well to antidepressant drugs, but a characteristic feature of the response is that there is a lag period of 7-14 days before the antidepressant effect becomes apparent. The most severely depressed patients who have psychotic symptoms often respond poorly to antidepressant drugs but recover rapidly with a course of electroconvulsive therapy (ECT).

Biochemical theories of the causes of depression have largely been derived from the action of drugs. Reserpine is the active alkaloid of the plant *Rauwolfia serpentina* and was used in the 1950s to treat schizophrenia and to lower blood pressure. It was noted that a subgroup of patients treated with reserpine developed depressive symptoms which were often severe. At about the same time, iproniazid, a monoamine oxidase (MAO) inhibitor, was used in the treatment of tuberculosis and was reported to produce euphoria and overactivity in a small group of patients. Reserpine acts by disrupting the granule storage of the monoamines noradrenaline, dopamine and 5-hydroxytryptamine (5-HT), thus depleting the stores of amine in the neurone (Fig. 2). The major effect of reserpine is, after a brief initial stimulation, to decrease the activity of noradrenaline, dopamine and 5-HT neurones. In animals, reserpine produces a "depression-like syndrome", including sedation and motor retardation, which is reversed by the precursor of dopamine and noradrenaline, L-DOPA, but not by L-tryptophan or 5-hydroxytryptophan, the precursors of 5-HT.

The first monoamine theory of affective illness was based on the actions of reserpine and iproniazid in respectively depleting and increasing (by reducing metabolism by MAO) monoamines in the brain. This theory suggested that an excess of monoamines led to the increased activity and elevation of mood characteristic of mania, and that a deficiency of monoamines resulted in depression. Since this theory was put forward in 1958 it has undergone modification and refinement, but it remains unproven, despite extensive investigation.

The major group or groups of drugs now used in the treatment of depression are the tricyclic antidepressants. These drugs are potent inhibitors of the sodium-dependent noradrenaline and 5-HT high-affinity uptake mechanism which rapidly terminates the action of these transmitters by re-uptake into the presynaptic neurone (Fig. 2). The commonly used antidepressants imipramine and amitriptyline are equally effective in blocking noradrenaline and 5-HT uptake, but more recent drugs have been developed which are selective towards either noradrenaline or 5-HT uptake blockade. Both selective noradrenaline-uptake inhibitors (desipramine, maprotiline and viloxazine) and selective 5-HT-uptake inhibitors (clomipramine, zimelidine, fluvoxamine and fluoxetine) have antidepressant properties. The acute effects of these drugs is to potentiate the action of noradrenaline and 5-HT, and although they have a number of other effects such as blockade of α_1 adrenoreceptors, muscarinic receptors and histamine receptors, these are less likely to be involved in their antidepressant action.

Complications arise from drugs with apparent antidepressant properties which have limited effects on monoamine re-uptake, for example, iprindole and mianserin. Mianserin is, however, a more effective antagonist at α_2 adrenoreceptors than other antidepressants and this will acutely increase noradrenaline release by inhibiting the feedback mechanism by which noradrenaline inhibits its own release through presynaptic α_2 receptors (Fig. 2).

More puzzling is the delay of at least 7-10 days before the antidepressant drugs show any therapeutic effect on the symptoms of depression. Blockade of noradrenaline and 5-HT uptake occurs within minutes after oral administration of the drugs. More recent research has focused on the chronic effects of antidepressant drugs and comparisons with the effects of electroconvulsive stimulation (ECS), the animal model of ECT [6]. The most consistent finding is the reduction in β-receptor-stimulated adenylate cyclase and the reduced density of β-adrenoreceptors which occurs in animals after at least 7 days' administration of any antidepressant drug [18]. The same "down-regulation" of β-receptors also occurs after seven to eight

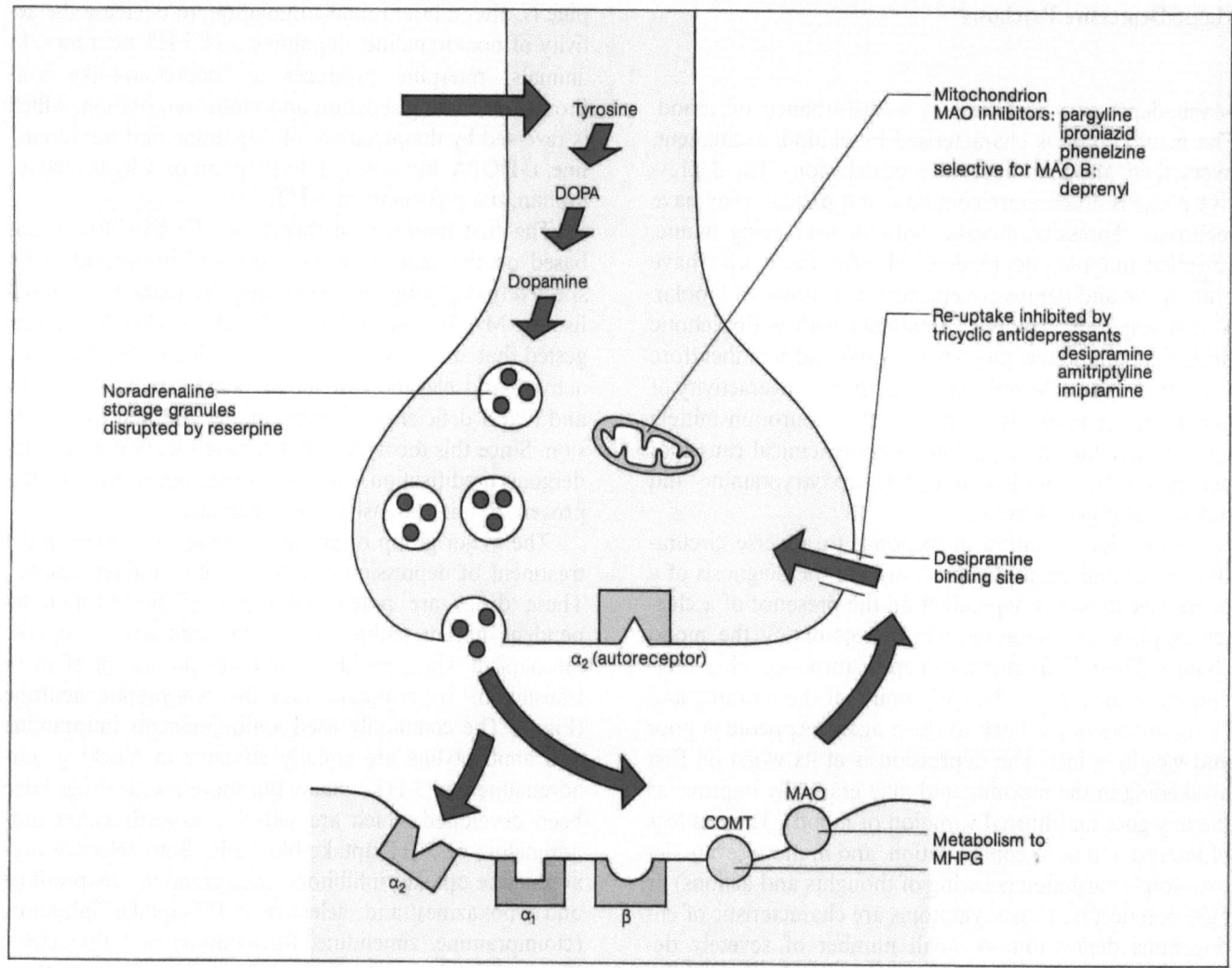

Fig. 2. Model of a noradrenaline synapse showing the principal sites of action of drugs which influence mood

ECS treatments given on successive days but not if the shocks are all given on the same day. This mimics the way in which ECT has to be given, normally two to three treatments a week, in order to have an antidepressant action. The changes in receptors after chronic antidepressant drugs are not confined to β-receptors, although other receptor changes are less consistent and vary with different antidepressants or ECS (Fig. 3). There is a less consistent "down-regulation" decrease of α_2-adrenoreceptors but α_1-adrenoreceptors are increased, "up-regulated". After chronic administration of some but not all antidepressant drugs there is "down-regulation" of 5-HT$_{1A}$-autoreceptors, and a course of ECS has the same effect [7]. The number of 5-HT$_2$ binding sites is reduced by chronic antidepressant administration, but ECS has the opposite effect, increasing the density of 5-HT$_2$ receptors and the behavioural syndrome associated with stimulation of these receptors. Thus, changes in both pre- and postsynaptic adrenergic and 5-HT receptors occur in approximately the time it takes for antidepressant drugs to have a therapeutic effect. However, only "down-regulation" of β-receptors and possibly α_2- and 5-HT$_{1A}$-receptors occurs with both antidepressants and ECS. It is difficult to interpret how these changes will affect the actual functioning of monoamine neurones in the normal human being, let alone how antidepressant drugs may modify the potentially abnormal function of these neurones or receptors in the depressed patient.

A variety of attempts have been made to assess the functioning of monoaminergic systems in depressed patients. Unfortunately, there is no way in which one can directly examine the biochemistry of the human brain, although the advent of positron emission tomography does offer some exciting possibilities. Studies of metabolites in urine and cerebrospinal fluid (CSF) and work on platelets have been productive but the relevance to what is hap-

Degenerative Disorders and Dementia

Parkinson's Disease

The first degenerative disease of the brain shown to be associated with the loss of a specific neurotransmitter was Parkinson's disease, but Huntington's chorea and Alzheimer-type dementia (ATD) have now been shown to also be associated with damage to specific neurotransmitter systems.

In Parkinson's disease there is characteristically a marked loss of dopamine in the substantia nigra and in almost all the forebrain structures innervated by dopamine (Fig. 1). There is degeneration of the dopamine cell bodies in the pars compacta of the substantia nigra, and before the symptoms of Parkinson's disease are apparent concentrations of dopamine in the caudate nucleus must be reduced by 75%–80% [9]. The symptoms of Parkinson's disease respond to treatment with L-DOPA, the precursor of dopamine. In the early stages of Parkinson's disease the symptoms (difficulty in initiation and movement, muscle rigidity and tremor) may be totally abolished by L-DOPA. It is usually used in combination with a peripheral DOPA decarboxylase inhibitor (carbidopa, benserazide), which increases the amount of L-DOPA available for decarboxylation to dopamine in the brain. Direct dopamine agonists - bromocriptine and lisuride, for example - are also useful in relieving the symptoms of Parkinson's disease.

Although the reduction of dopamine is the most severe neurotransmitter deficit in Parkinson's disease, there are changes in other neurotransmitter systems as well. A proportion of mesolimbic dopamine neurones contain cholecystokinin, and this peptide is reduced in Parkinson's disease. In the nigrostriatal system there is a reduction of 5-HT and of the GABA synthetic enzyme, glutamate decarboxylase (GAD). Noradrenaline is reduced in the substantia nigra and the nucleus accumbens, and there is cell loss in the locus coeruleus, the origin of the noradrenaline projection to the cortex. Choline acetyltransferase (CAT) is also reduced in some parkinsonian brains and there may be cholinergic cell loss in the basal forebrain. The nigrostriatal dopamine neurones have an inhibitory effect on cholinergic striatal interneurones; thus, anticholinergic drugs have a modest effect in reducing the symptoms of Parkinson's disease.

The therapeutic action of L-DOPA in Parkinson's disease may become reduced with the progression of the disease, and approximately half of the patients treated for several years with L-DOPA will develop "on-off" effects. This presents as the sudden and unpredictable onset of immobility with the patient describing himself as freezing or getting stuck. The cause of these sudden fluctuations in mobility remains uncertain, but they tend to occur after peak plasma L-DOPA concentrations have been reached and may be related to a "wearing-off" reaction. Frequent administration of small doses of L-DOPA or its combina-

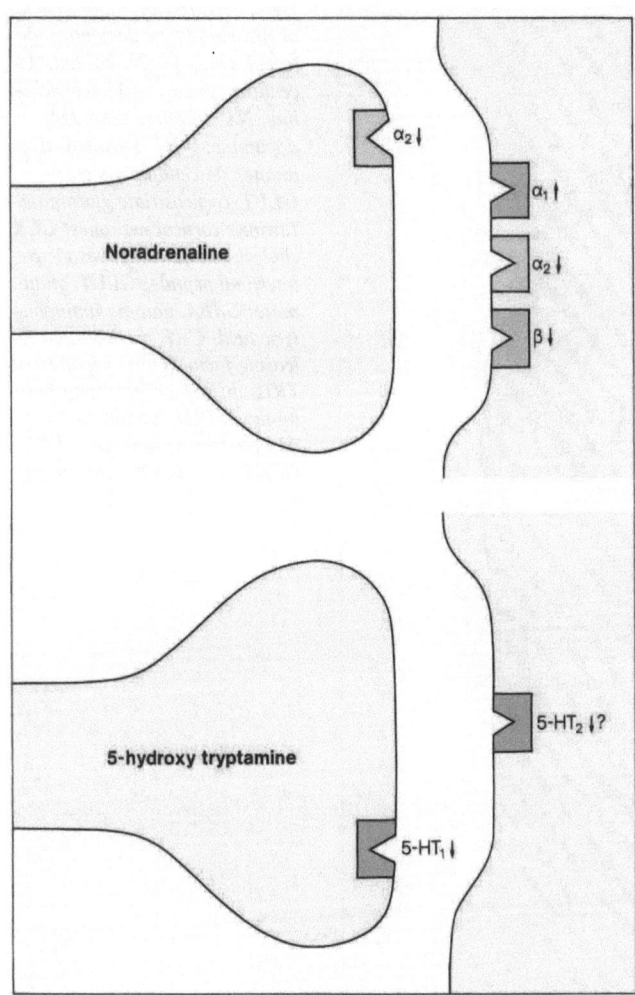

Fig. 3. Changes in noradrenaline and 5-hydroxytryptamine receptors which occur after chronic administration of antidepressant drugs

pening in the brain remains uncertain. Neuroendocrine tests bring research a step nearer to understanding the brain, the output of hormones from the anterior pituitary being under the direct control of a variety of neurotransmitters, including noradrenaline, dopamine and 5-HT. The results of neuroendocrine challenge studies suggest that depressed patients have decreased hormonal responses to drugs which enhance noradrenergic function, for example, the growth hormone response to the α_2-agonist clonidine, and that changes in hormonal responses occur after the administration of antidepressant drugs indicative of the receptor changes that have been shown to occur in animal studies [16] (see pp. 55 ff. for details).

Thus, although noradrenaline and 5-HT systems are still thought to be involved in some way in the cause of depressive illnesses, despite considerable research effort we remain ignorant as to the specific abnormalities of function.

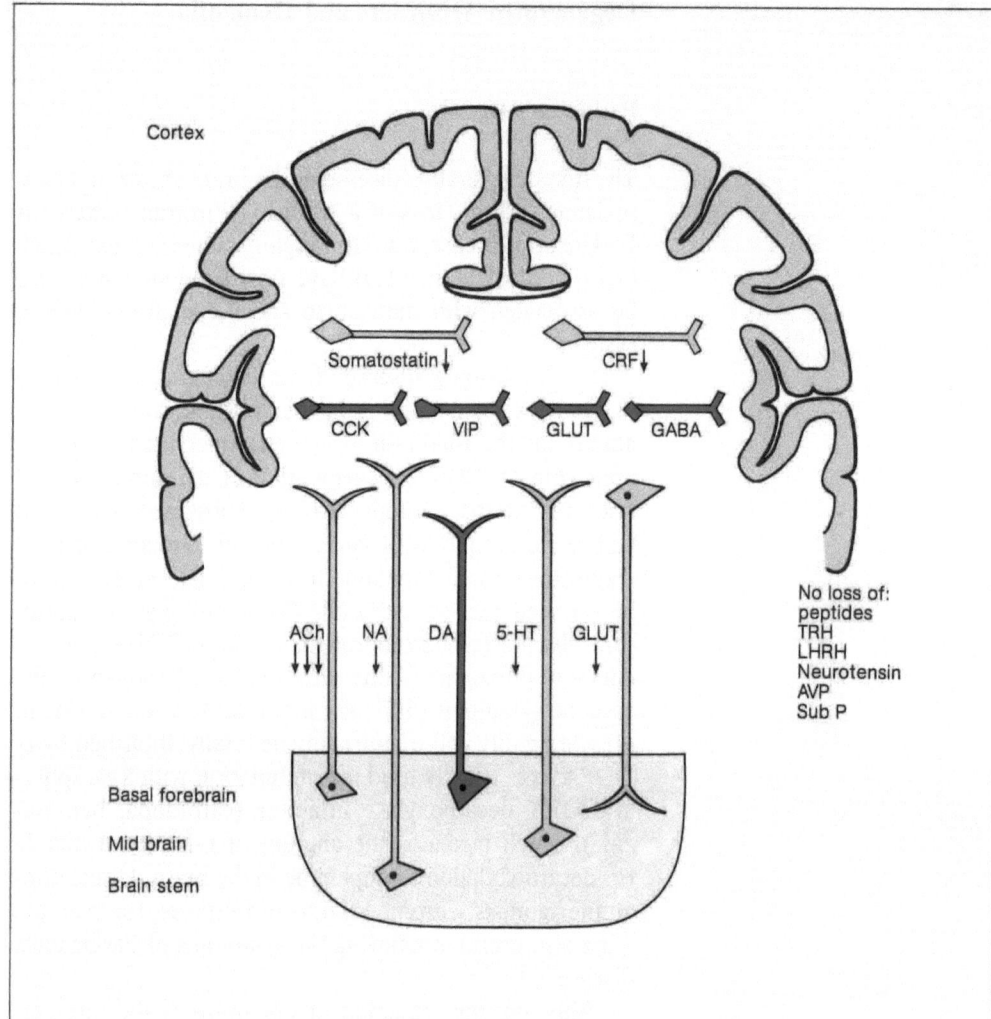

Fig. 4. Neurotransmitter systems in Alzheimer-type dementia: affected, blue; unaffected, red. Ascending systems: **Ach**, acetylcholine; **NA**, noradrenaline; **DA**, dopamine; **5-HT**, 5-hydroxytryptamine. Descending system: **GLUT**, corticostriate glutamate. Intrinsic cortical neurones: **CCK**, cholycystokinin; **VIP**, vasoactive intestinal peptide; **GLUT**, glutamate; **GABA**, gamma aminobutyric acid; **CRF**, corticotropin-releasing factor. Other peptides: **TRH**, thyrotropin-releasing hormone; **LHRH**, luteinising hormone-releasing hormone; **AVP**, vasopressin; **Sub P**, substance P

tion with the type-B MAO inhibitor deprenyl may reduce the "on-off" effects, and most recently, constant infusion of L-DOPA by minipump has proved effective. Mental deterioration with the development of a global dementia is also a common complication of late Parkinson's disease.

Alzheimer-type Dementia

Dementia is a global deterioration in intellectual function, and the most common cause of dementia is ATD. Approximately 5% of people aged 65 years have some degree of dementia; for people aged 80 years this figure rises to 20%, and more than 50% of dementia patients have ATD. The initial symptom of ATD is memory impairment for recent events with difficulty in retension of new information. There is progressive impairment of intellectual functions, with speech difficulties (dysphasia) and signs of parietal lobe dysfunction (dyspraxia). The neuropathological changes in ATD are general shrinkage of the cortical gyri with enlargement of the sulci, and there are characteristic microscopic features: senile plaques and intraneuronal neurofibrillary tangles [19].

In ATD there is selective loss of neurotransmitters in postmortem brain (Fig. 4). The major loss is in acetylcholine, but there are also reductions in noradrenaline, 5-HT and somatostatin [8]. The synthetic enzyme for acetylcholine, CAT, is reduced in most brain areas, particularly in the hippocampus where it is reduced to approximately 20% of control levels. There is also a reduction in the synthesis of acetylcholine in brain biopsy tissue from the temporal cortex of ATD patients. The loss of cholinergic markers in the cortex results from damage to the ascending projection of cholinergic neurones from the basal forebrain. The majority of studies have shown no reduction of muscarinic receptor binding in ATD.

The ascending noradrenaline systems are also involved in ATD, with loss of pigmented cells from the locus coeruleus, but the reduction in the noradrenaline synthetic enzyme dopamine-β-hydroxylase is less consistent than the loss of CAT. There is also evidence for involvement of the ascending 5-HT projections from the raphe nuclei and the descending corticostriatal glutamate pathway in ATD [2]. To date, the only intrinsic cortical neurotransmitters shown to be affected in ATD are the peptides somatostatin, corticotropin-releasing factor, and possibly neuropeptide Y; the concentrations of other peptides are normal (Fig. 4) [5]. It remains an area of speculation whether the cortical changes in ATD (senile plaques and neurofibrillary tangles) are secondary to damage to the ascending cholinergic and monoaminergic systems or whether the cortical changes occur first and there is secondary damage to the ascending systems [14].

The neuropathological changes and many of the neurochemical changes of ATD also occur in Down's syndrome (trisomy 21) and are present in the brain of all Down's patients who die after the age of 40 years [21].

The success of L-DOPA therapy in Parkinson's disease prompted the evaluation of cholinergic drugs in ATD. The effects of the precursors of acetylcholine, choline and lecithin, have been extensively studied in ATD but they appear to have no beneficial effects [1]. Physostigmine, a cholinesterase inhibitor which reduces acetylcholine metabolism, can improve memory in normal subjects and also produces a limited improvement in performance or memory tests in ATD patients, as does the muscarinic agonist arecoline [3]. The positive effects of these drugs are slight, however, and they do not offer a practical therapy for ATD.

Huntington's Disease

Huntington's chorea is a dominantly inherited disease characterised by abnormal movements (choreiform) and dementia. Family studies show that the gene for Huntington's disease is linked to a polymorphic DNA marker that maps to human chromosome 4 [8]. The basal ganglia is primarily involved and there is extensive degeneration of the striatum. The major neurotransmitter deficits in Huntington's disease are in striatal GABA and in cholinergic interneurones [17]. Dopamine neurones are not damaged, and there is a relative overactivity of the nigrostriatal dopamine system which causes involuntary movements similar to those seen after excessive dosage of L-DOPA. Despite the dementia that occurs in Huntington's disease, neurochemical studies show that there is essentially no evidence for involvement of the ascending neurotransmitter systems which are characteristicly involved in ATD.

Conclusions

The 'dopamine' hypothesis of schizophrenia and various modifications of the 'monoamine' hypothesis of affective disorders remain to be disproved, but clear consistent neurochemical abnormalities, independent of drug effects, have yet to be demonstrated. Degenerative disorders of the central nervous system, including Parkinson's disease, Huntington's disease and ATD, are associated with different patterns of loss of specific neurotransmitters, but only in Parkinson's disease has replacement therapy in the form of L-DOPA proved effective.

References

1. Bartus RT, Dean RL, Beer B, Lippa AS (1982) The cholinergic hypothesis of geriatric memory dysfunction. Science 217: 408-417
2. Bowen DM, Davison AN, Francis PT, Palmer AM, Pearce BR (1985) Neurotransmitter and metabolic dysfunction in Alzheimer's dementia: relationship to histopathological features. In: Rose FC (ed) Modern approaches to the dementias; Interdisciplinary topics in gerontology, vol 19. Karger, Basel, pp 156-174
3. Christie JE, Shering A, Ferguson J, Glen AIM (1981) Physostigmine and arecoline: effects of intravenous infusions in Alzheimer presenile dementia. Br J Psychiatry 138: 46-50
4. Crow TJ (1980) Molecular pathology of schizophrenia: more than one disease process? Br Med J 280: 1-9
5. De Souza EB, Whitehouse PJ, Kuhar MJ, Price DL, Vale WW (1986) Reciprocal changes in corticotropin-releasing factor (CRF)-like immunoreactivity and CRF receptors in cerebral cortex of Alzheimer's disease. Nature 319: 593-595
6. Garattini S, Samanin R (1984) Drugs: guide and caveats to explanatory and descriptive approaches. I. A critical evaluation of the current status of antidepressant drugs. J Psychiatr Res 18: 373-390
7. Goodwin GM, De Souza RJ, Green AR (1985) Presynaptic serotonin receptor-mediated response in mice attenuated by antidepressant drugs and electroconvulsive shock. Nature 317: 531-533
8. Gusella JF, Wexler NS, Conneally M, Naylor SL, Anderson MA, Tanzi RE, Watkins PC, Ottina K, Wallace MR, Sakaguchi AY, Young AB, Shoulson I, Bonilla E, Martin JB (1983) A polymorphic DNA marker genetically linked to Huntington's disease. Nature 306: 234-238
9. Hardy J, Adolfsson R, Alafuzoff I, Bucht G, Marcusson J, Nyberg P, Perdahl E, Wester P, Winblad B (1985) Transmitter deficits in Alzheimer's disease. Neurochem Int 7: 545-563
10. Hornykiewicz O (1979) Brain dopamine in Parkinson's disease and other neurological disturbances. In: Horn AS, Korf J, Westerink BHC (eds) The neurobiology of dopamine. Academic, London, pp 633-654
11. Johnstone EC, Crow TJ, Frith CD, Stevens M, Kreel L, Husband J (1978) The dementia of dementia praecox. Acta Psychiatr Scand 57: 305-324
12. Mackay A, Iversen L, Rossor M, Spokes E, Bird E, Arregui A, Creese I, Snyder S (1982) Increased brain dopamine and dopamine receptors in schizophrenia. Arch Gen Psychiatry 39: 991-997
13. Reynolds GP (1983) Increased concentrations and lateral asymmetry of amygdala dopamine in schizophrenia. Nature 305: 527-529
14. Rossor MN, Mountjoy CQ, Roth M, Reynolds GP (1985) Ascending systems in Alzheimer's disease. In: Rose FC (ed) Modern approaches to the dementias; Interdisciplinary topics in gerontology, vol 19. Karger, Basel, pp 198-212
15. Roth M (1986) The association of clinical and neurological findings and its bearing on the classification and aetiology of Alzheimer's disease. Br Med Bull 42: 42-50
16. Siever LJ, Uhde TW (1984) New studies and perspectives on the noradrenergic receptor system in depression: effects of the α_2-adrenergic agonist clonidine. Biol Psychiatry 19: 131-156
17. Spokes EGS (1980) Neurochemical alterations in Huntington's chorea. A study of post-mortem brain tissue. Brain 103: 179-210
18. Sulser F (1983) Deamplification of noradrenergic signal transfer by antidepressants; a unified catecholamine-serotonin hypothesis of affective disorders. Psychopharmacol Bull 19: 300-304
19. Terry RD, Katzman R (1983) Senile dementia of the Alzheimer type. Ann Neurol 14: 497-506
20. Weinberger DR (1984) Computed tomography (CT) findings in schizophrenia: speculation on the meaning of it all. J Psychiatr Res 18: 477-490

21. Yates CM, Simpson J, Gordon A, Maloney AFJ, Allison Y, Ritchie IM, Urquhart A (1983) Catecholamines and cholinergic enzymes in pre-senile and senile Alzheimer-type dementia and Down's syndrome. Brain Res 280: 119–126

Normal and Disordered Central Neurotransmitter Function Studied through the Neuroendocrine Window of the Brain

George Fink

MRC Brain Metabolism Unit, University of Edinburgh, Department of Pharmacology, Edinburgh, United Kingdom

The preceding chapter reviewed the pharmacological evidence which shows that disorders of movement, memory, mood and thought appear to be due to abnormalities in central chemical neurotransmission. However, access to the brain to obtain direct information on central neurotransmission at physiological sites (i.e. synapses) and under physiological conditions is difficult to achieve. As mentioned in previous chapters, the two main criteria for a neurotransmitter are, first, that the substance is released from nerve terminals in which it is located and, secondly, that its actions on the appropriate effector cells are compatible with the proposed function of the transmitter. These two criteria are easy to satisfy in the autonomic and peripheral nervous systems where nerve terminals, synapses and effector cells are readily accessible to investigation and experimental manipulation, but they are difficult to satisfy in the central nervous system. However, the brain does control the secretion of pituitary hormones. The neuroendocrine neurones which comprise the final common pathway[1] for the neural control of the pituitary gland are located in the hypothalamus. The hypothalamic neuroendocrine neurones are comprised of large cells (the magnocellular neurones) that terminate on and release their neurotransmitters into systemic blood vessels of the neural lobe of the pituitary gland (the hypothalamo-neurohypophysial system), and smaller neurones (parvocellular) that terminate on the primary plexus of the hypophysial portal vessels which convey transmitters (or 'neurohormones' to be precise) released to the anterior pituitary gland (the hypothalamo-adenohypophysial system), where they either facilitate or inhibit the secretion of anterior pituitary hormones (Fig. 1). The activity of the neuroendocrine neurones is greatly affected by the activity of inputs from other regions of the brain, especially the limbic system, which is involved in emotion and responses to stressful stimuli. Thus, the hypothalamic-pituitary system provides a neuroendocrine window on central neurotransmission by means of which it is possible to characterize central neurotransmitter function either directly by measuring neurotransmitters in hypophysial portal blood or indirectly by measuring the output of pituitary hormones into the systemic circulation. The hypophysial portal vessels can, in fact, be considered as a neurovascular synapse. As shown by electron microscopy and immunocytochemistry, the nerve terminals which terminate on the primary plexus of the portal vessels are heterogeneous (Fig. 2), and this carries the advantage that it is possible to

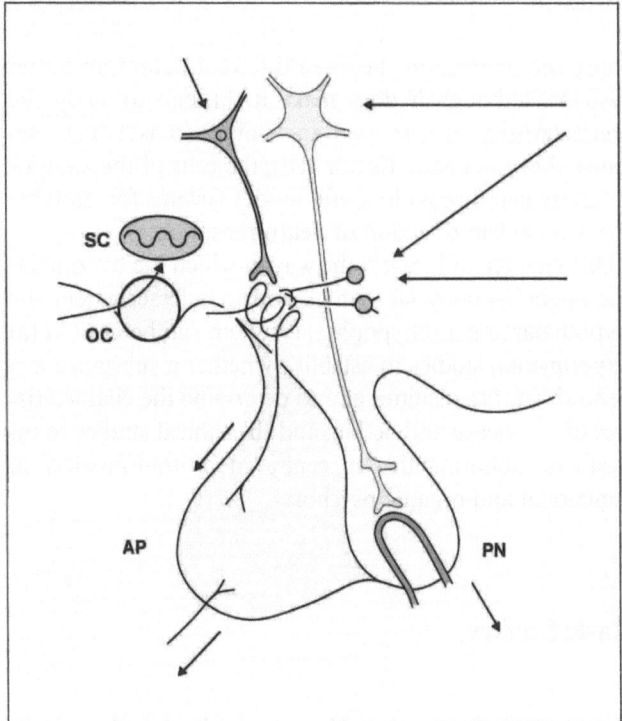

Fig. 1. The hypothalamic-pituitary system, showing the magnocellular (**yellow**) projections directly to the systemic vessels of the pars nervosa (**PN**) and the parvocellular (**green**) projections to the primary plexus of the hypophysial portal vessels, which convey transmitters to the pars distalis of the anterior pituitary gland (**AP**). Dorsal to the optic chiasm (**OC**) are the suprachiasmatic nuclei (**SC**), which receive direct projections from the retina and play a key role in the control of circadian and other rhythms (indicated by the sinusoidal curve). The activity of the intrinsic neurones of the hypothalamus is greatly influenced by projections (**arrows**) from numerous areas of the forebrain and hindbrain, particularly the limbic system, as well as by hormones, mainly oestrogen, progesterone and prolactin, in the case of the hypothalamic-pituitary-gonadal system

---

[1] G.W. Harris borrowed this term from Sherrington, who first applied it to the α motor neurones of the spinal cord [42].

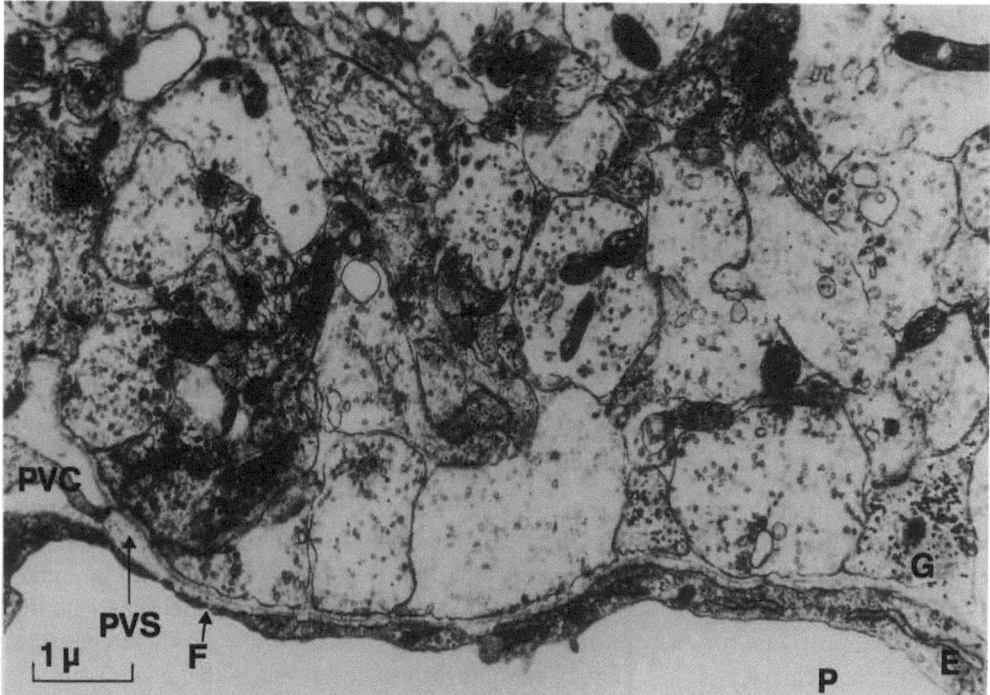

Fig. 2. Section taken through the median eminence of the rat viewed under the electron microscope. The electron micrograph shows part of a primary capillary (**P**) of the hypophysial portal system surrounded by a perivascular space (**PVS**) of about 100 nm. The wall of the capillary has fenestrations (**F**) and is surrounded by nerve and glial cell (**G**) terminals. Transmitters released at the various types of nerve terminals must diffuse through the PVS to reach the portal vessel. The process of a perivascular cell (**PVC**) may be seen. (From [34], reproduced with permission)

study the interactions between different neurotransmitter systems, although it does make it difficult to study the characteristics of one neurotransmitter in isolation. Because they are neuroeffector cells, the cells of the anterior pituitary gland provide useful model systems for studying the mechanism of action of neurotransmitters.

This chapter will review the way in which the hypothalamo-adenohypophysial system, and to a lesser extent the hypothalamo-neurohypophysial system can be used in (a) experimental studies to establish whether a substance is a central neurotransmitter and to determine the characteristics of its release and action, and (b) clinical studies to investigate abnormalities in central neurotransmission in functional and organic psychoses.

Basic Studies

Somatostatin Release Into Hypophysial Portal Blood: Determination of Whether a Substance is a Central Neurotransmitter

Somatostatin 28 provides an example of how the measurement of neuropeptide release into hypophysial portal blood can be used to determine whether a peptide is a central neurotransmitter. Somatostatin (SS) was first isolated from the hypothalamus as a 14 amino acid peptide transmitter which inhibits growth hormone secretion [9]. Recombinant DNA technology showed that, as for other neuropeptides, somatostatin is synthesized in the neurone as part of a large precursor polypeptide [38, 88]. Larger forms of somatostatin, in particular SS 28, were also isolated [73], and since SS 28 was also found to inhibit the release of growth hormone the question arises whether SS 28 is a neurotransmitter in its own right. This question can be answered by determining whether SS 28 is released into hypophysial portal blood.

As assessed by radioimmunoassay the concentrations of SS in hypophysial portal vessel blood are about ten fold higher than those in the peripheral circulation [61]. Stimulation of the hypothalamus results in a relatively massive increase in the output of SS into the portal vessels (Fig. 3). Analysis of the SS concentration in hypophysial portal vessel blood with the aid of high-performance liquid chromatography shows that both SS 28 and SS 14 are released into the portal vessels from nerve terminals in the hypothalamus (Fig. 4). Furthermore, studies carried out with the aid of an antibody specific to the N-terminus of SS 28 show that SS 28 (1-12) is also released into portal vessel blood in concentrations higher than those in peripheral blood and that the output of this peptide can be further increased by electrical stimulation of the hypothalamus [90]. These results show that SS 28 is released as a neurotransmitter in its own right and provide important clues for our understanding of the physiology of the control of growth hormone secretion and the development of clinically useful analogues of both SS 14 and SS 28.

The results of these studies on hypophysial portal blood, taken together with those obtained by measuring SS 14 and SS 28 release from hypothalamic and ME slices in vitro [41], suggest that the SS precursor is processed differently at different sites. A likely, but obviously not the only explanation for the release of both SS 28 and SS 14 from the ME [41] is that there are two different types of

Fig. 3. The mean (± SEM) concentrations of immunoreactive somatostatin per ml of plasma and the total amount released into hypophysial portal plasma collected from six animals during four consecutive periods of 15 min. An electrical stimulus was applied to the median eminence during the second and fourth periods of collection (**Stim**). Using the same extraction and assay procedures, the amount of somatostatin detected in blood from the inferior vena cava and aorta was 35 pg/ml. (From [61], with permission)

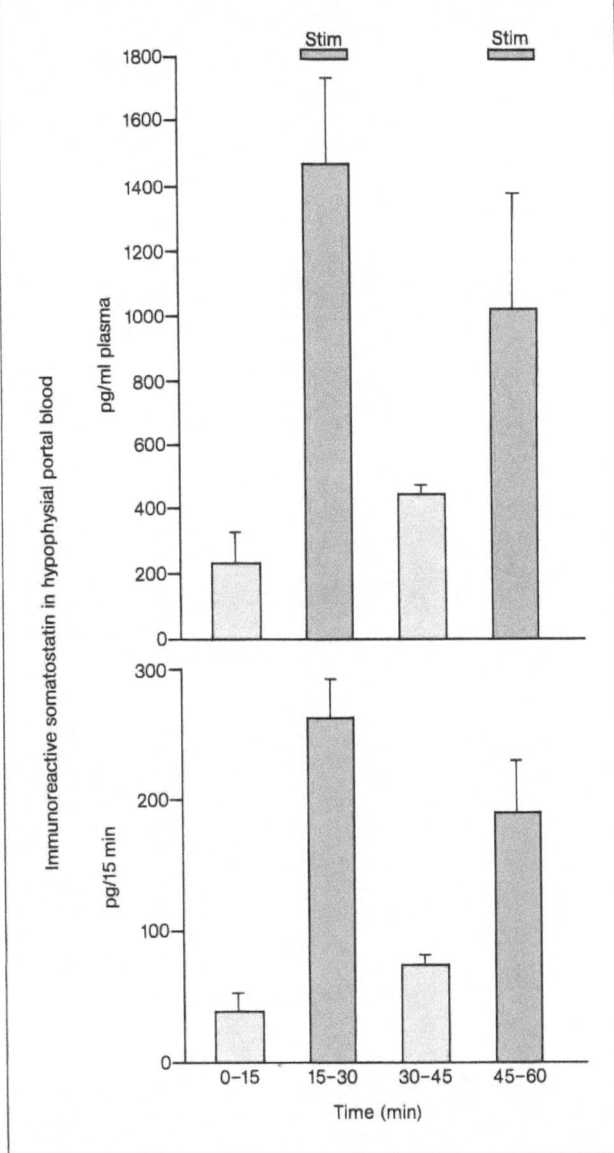

Fig. 4. Representative profiles of somatostatin immunoreactivity in HPLC fractions of extracts of portal plasma collected during (**left**) and after (**right**) the application of an electrical stimulus to the median eminence, showing that two peaks of somatostatin activity are present in portal plasma, one of which corresponds to somatostatin 14 and the other to somatostatin 28. (From [61], with permission)
▼

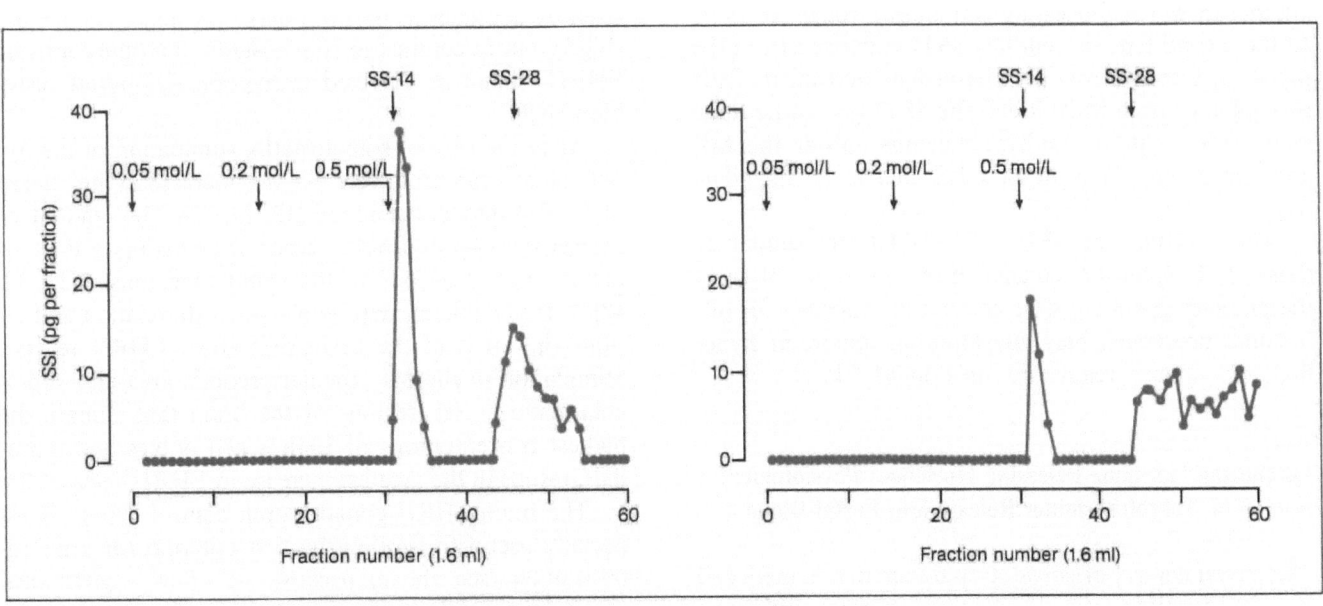

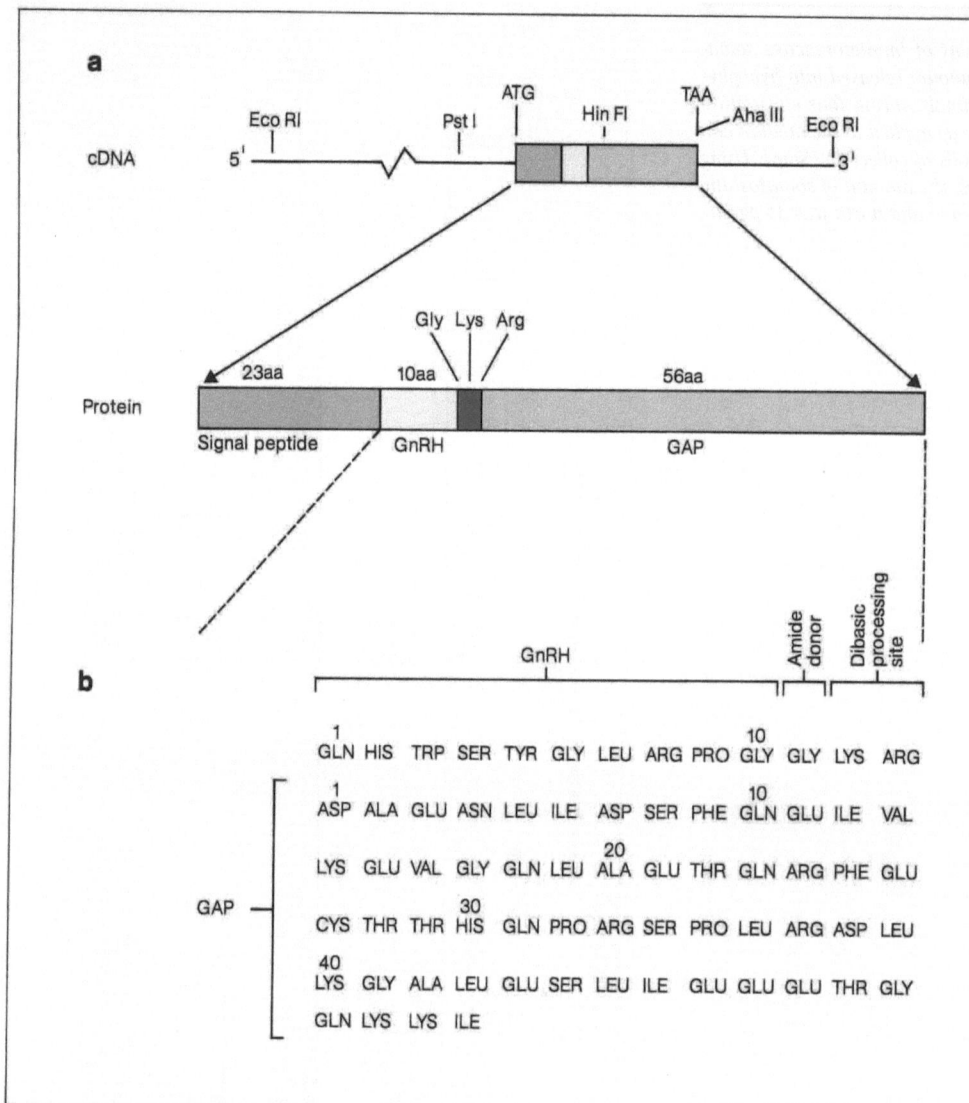

Fig. 5a, b. Complementary DNA and precursor protein for human LHRH; structure and encoded amino-acid sequence of human placental cDNA for prepro-LHRH. a Partial restriction endonuclease map. The coding region is located between the initiation codon for protein synthesis ATG and the termination codon TAA. Schematic representation of the encoded protein identifies the three domains of signal peptide, gonadotropin-releasing hormone (GnRH) (synonym for LHRH) and gonadotropin-releasing hormone-associated peptide (GAP) with the respective sizes in amino-acid (aa) residues. b Amino-acid sequences of LHRH and GAP with an enzymatic processing site separation the two moieties. Numbers refer to the respective positions within LHRH(1-10) or GAP(1-56). (From [66], with permission of the authors and Macmillan Journals Ltd)

SS neurones: one in which processing stops at SS28 and another in which processing continues through to SS14. In the second type of neurone, SS14 is derived from the precursor directly, from an intermediate derivative of the precursor, or from SS28 itself. The SS14 type of neurone may predominate in the hypothalamus outside the ME and also in other areas of the CNS such as the amygdala [71].

The functions of SS28 are qualitatively similar to those of SS14, but the potency of the two forms of SS in the pituitary gland are different and the capacity for differential processing may constitute an important hypothalamic-pituitary regulatory function [61, 71].

Luteinizing Hormone-Releasing Hormone: Physiological Studies of Neurotransmitter Release into Portal Blood

The neural control of gonadotropin secretion is mediated by a decapeptide termed 'luteinizing hormone-releasing hormone' (LHRH). Like somatostatin, it too is synthesized by neurones as part of a large precursor [19, 66, 87] (Fig. 5), but in contrast to somatostatin, multiple forms of LHRH cannot be detected in hypophysial portal vessel blood [92].

As in the case of somatostatin, stimulation of the hypothalamus can produce a massive increase in the release of LHRH into portal blood [12, 24, 32]. The amount of LHRH released into portal blood depends upon the amplitude and frequency of the stimulating pulses [23, 32, 48]. It is possible to map by electrical stimulation and lesions the areas of the brain that affect LHRH release. Stimulation of the ME, medial preoptic area and suprachiasmatic nuclei, regions of the brain that contain the highest concentrations of LHRH cell bodies and terminals, results in the greatest increase in LHRH release [12].

The brain-LHRH-gonadotropin control system is especially useful for testing the first criterion for a neurotransmitter (see above) because we know a great deal about the physiology of gonadotropin secretion. Thus, the advantages of studying LHRH release are that (a) the release of the peptide and its action on the anterior pituitary gland can be correlated with the plasma concentrations of LH, the pattern of which has been well defined; (b) the brain areas that control LH release have been relatively well defined, and this provides a rational basis for studying the effects of electrical stimulation of the brain on LHRH release (see above); (c) the several patterns of LH release that occur spontaneously and that are induced experimentally by steroids provide useful models for studying LHRH release generated by endogenous signals; and (d) pharmacological studies have shown that monoaminergic and opioid neurones control or influence LHRH neurones and, therefore, measurement of LHRH release can provide information on the actions of central monoaminergic and opioid neurones and the interactions between non-peptidergic and peptidergic neurones.

The Spontaneous Surge of LHRH

A massive surge of LH, which causes ovulation, occurs once during each oestrous or menstrual cycle in spontaneously ovulating mammals. The spontaneous surge of LH in both rats and monkeys is triggered by a surge of LHRH into hypophysial portal blood [13, 64, 80, 89]; this in turn is triggered by the surge of oestradiol-17β, which precedes the LHRH surge [65, 81, 82]. Figure 6 shows the salient features of the positive feedback cascade that leads to the spontaneous LH surge. The spontaneous LHRH surge is relatively small in both the rat and the monkey; concentrations increase from basal values of about 20-30 pg LHRH/ml to 100-200 pg LHRH/ml of portal plasma. Studies in which LHRH was infused intravenously into rats at dioestrus and pro-oestrus showed that the concentrations of LHRH reached at the peak of the surge can produce a surge of LH only in animals (pro-oestrus) in which the responsiveness of the pituitary gland to LHRH has increased 20- to 50-fold [29]. A 20- to 50-fold increase in pituitary responsiveness to LHRH occurs in the rat between dioestrus and the late afternoon of pro-oestrus [3] and between the early follicular phase and the ovulatory phase of the menstrual cycle in human beings [111].

While the LHRH surge is probably important for ensuring the precise timing of the spontaneous LH surge, it must be stressed that exposure of the pituitary gland to small pulses of LHRH [26, 30] or continuous infusion of small amounts of LHRH [98] can also produce a massive surge of LH, probably by way of the priming effect of LHRH (see below).

Pulsatile Release of LHRH

For most of the oestrous and menstrual cycles the secretion of LH is kept low by the negative feedback actions of oestrogen and progesterone. This basal secretion is not steady, but occurs in the form of pulses. The LH pulses are much more prominent when the feedback loop is opened by removal of the gonads, and under these conditions pulsatile release of LHRH into portal blood can be detected in ovariectomized rats [83], rhesus monkeys [11] and sheep [17]. The amplitude of LHRH pulses is related to the mean plasma LH concentrations [83], and in the rat, but not in the rhesus monkey, the pulsatile release of LHRH can be rapidly reduced by the intravenous injection of oestradiol-17β. Pulsatile LH release also occurs in intact rats, monkeys and human beings. Evidence from measurements of LHRH in portal blood (see above) and the effects of injecting LHRH or anti-LHRH sera (immunoneutralization) shows that the pulses of LH in intact animals are due to pulses of LHRH [27, 53].

The functional significance of the pulsatile release of LHRH, and as a consequence LH, is that (a) it provides the hypothalamic-pituitary-ovarian regulatory system with the capacity of control by both frequency and amplitude modulation, (b) small frequent pulses of LHRH could, by way of the priming effect of LHRH, lead to a spontaneous ovulatory surge of LH [30], and (c) pulsatile LHRH release prevents the down-regulation of LHRH receptors which occurs during continuous exposure to high levels of LHRH and leads to pituitary refractoriness to LHRH [53, 63, 86].

In the female rat and human being changes in both pulse frequency and amplitude are important in the signalling system [35, 112], but in the male rhesus monkey changes in the pulse frequency alone [72] may play a major role in the negative feedback control of LH release. That is, testosterone exerts its inhibitory effects on the hypothalamic-pituitary-gonadotropin system by reducing LH pulse frequency.

Circadian LHRH Release

Long-term ovariectomized rats treated with oestrogen alone or with oestrogen and progesterone have been used to provide several different experimental models for investigating the way in which steroids control the output of the hypothalamic-pituitary-gonadotropin system [28]. Long-term ovariectomized rats treated with high doses of oestradiol for several days show diurnal surges of LH which are due to diurnal surges of LHRH [83]. However, as in the case of the spontaneous surge of LH, the diurnal surges of LHRH are small and produce surges of LH only because the responsiveness of the anterior pituitary gland to LHRH in ovariectomized rats treated with oestrogen is more than two orders of magnitude greater than the level of pituitary responsiveness at dioestrus [83]. The diurnal surges of LHRH and LH in long-term ovariectomized rats treated with oestrogen provide strong support for the occurrence of a daily neural signal for LH release which is expressed in the form of an LHRH/LH surge only when the hypothalamic-pituitary system is exposed to high levels of oestradiol [25, 56, 83].

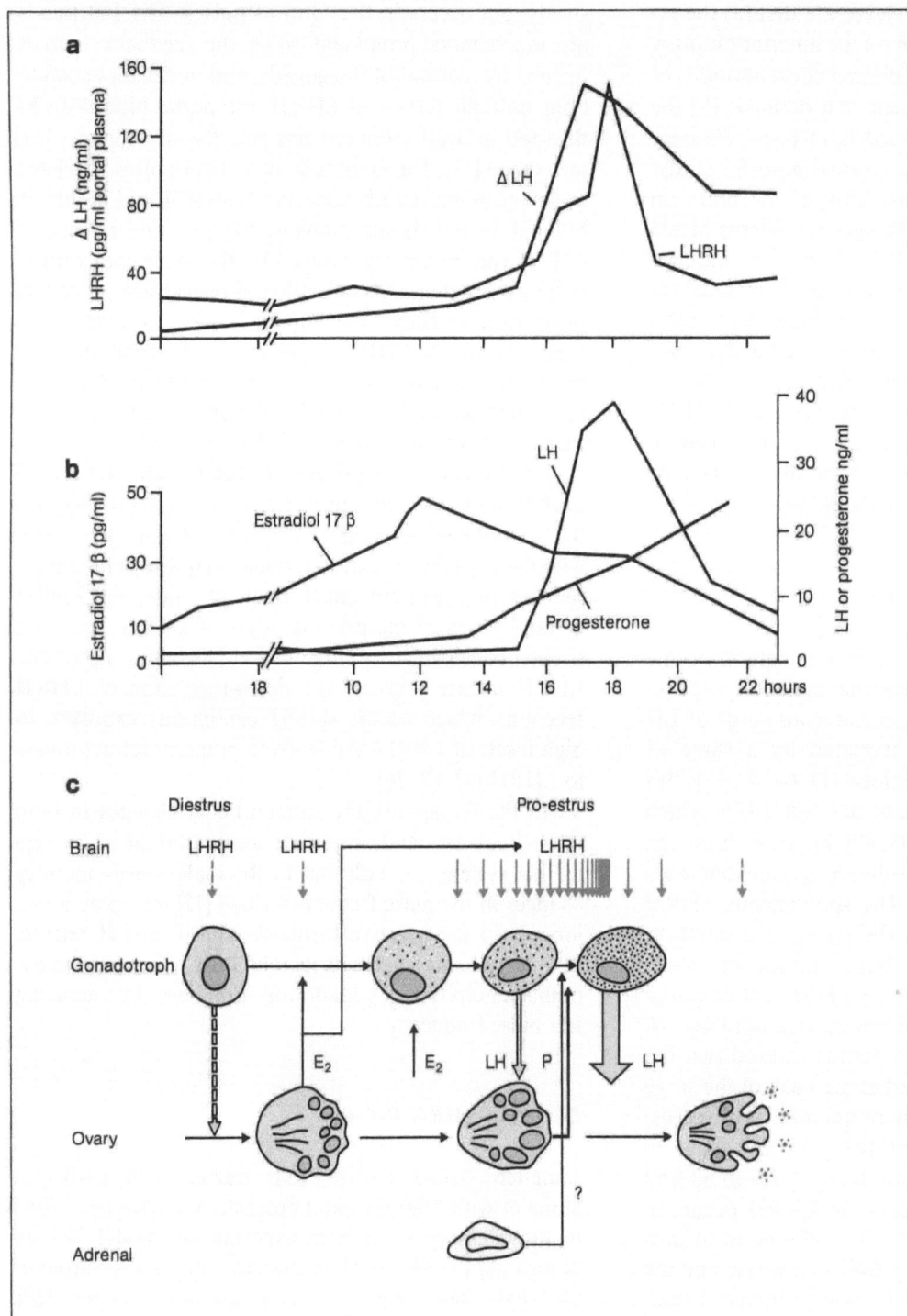

Fig. 6a-c. Cascade of events leading to the spontaneous ovulatory surge in the rat. *a* Changes in pituitary responsiveness (*ΔLH*) and mean concentrations of luteinizing hormone-releasing hormone (*LHRH*) in hypophysial portal plasma during dioestrus and pro-oestrus. *ΔLH*: mean maximal increments in peripheral plasma LH after intravenous injection of 50 ng LHRH/100 g body wt. *b* Mean peripheral plasma concentrations of oestradiol-17β (E_2), progesterone and LH. *c* Schematic diagram shows that plasma E_2 concentrations increase significantly before the LH surge: this ovarian signal increases the responsiveness of the pituitary gonadotropes (*increased stippling*) to LHRH and also triggers the surge of LHRH. Pituitary responsiveness to LHRH is further increased by progesterone secreted from the ovary in response to the LH released during the early part of the LH surge and by the priming effect of LHRH. The priming effect of LHRH co-ordinates the surge of LHRH with increasing pituitary responsiveness, so that the two events reach a peak at the same time. The conditions are thereby made optimal for a massive surge of LH. This cascade, which represents a form of positive feedback, is terminated by the rupture of the ovarian follicles (ovulation)

Site and Mechanism of Action of Oestradiol

Broadly, oestrogen has two major effects on the LHRH/LH release system: low plasma concentrations of oestrogen inhibit (negative feedback) while high plasma concentrations of oestradiol stimulate (so-called positive feedback) LHRH release. The negative feedback action is so fast that it may not necessarily involve protein synthesis: conceivably oestrogen could inhibit LHRH by a direct action on the membranes, ion channels, and enzymes involved in LHRH release. In contrast, the stimulatory action of oestrogen on LHRH release takes 26-28 h, and this is more than sufficient time to allow for protein synthesis and structural changes in the cytoskeleton, processes and synapses of neurones. Such an action would presumably involve protein synthesis, and indeed blockade of RNA synthesis and protein synthesis [46, 47] blocks the stimulation of LH release by oestrogen. However, combined immunocytochemical and autoradiographic studies have shown that only a few (one in 435) hypothalamic LHRH neurones contain nuclear oestrogen receptors, and, therefore, the effects of oestrogen that involve a genomic mechanism must be mediated by neurones that do not themselves contain LHRH but that terminate on LHRH neurones [94]. The likely candidates are noradrenergic (NA) neurones which stimulate LHRH and LH release [5, 31, 84] and dopaminergic (DA) and opioid neurones that inhibit LHRH release [5, 31, 36, 51, 84]. Figure 7 illustrates schematically how oestrogen could conceivably stimulate LHRH release by facilitating the activity of NA neurones and/or inhibiting the activity of DA or opioid neurones. The role of serotonergic neurones is unclear, but they may play a role in the daily signal for LHRH release since there is a dense serotonergic innervation of the suprachiasmatic nuclei, and blockade of serotonin synthesis blocks the diurnal surge of LH (in long-term ovariectomized rats treated with oestrogen) as well as the spontaneous surge of LH [43].

Luteinizing Hormone-Releasing Hormone: Possible Neurotransmitter Involved in Mating of Rats and Bullfrog Ganglia

LHRH was shown early on to potentiate lordosis in ovariectomized and hypophysectomized rats, and this suggested that the decapeptide was an important transmitter in mating behaviour [95]. Infusion studies of LHRH agonists and antagonists and anti-LHRH sera suggest that this action of LHRH may be located in the midbrain central grey [79, 96], which also contains LHRH nerve terminals [94] and shows increased uptake of [^{14}C]2-deoxyglucose in female voles who showed a lordosis response to males [99]. As assessed by immunocytochemistry, the apparent LHRH content of the LHRH fibres in the midbrain is influenced by gonadectomy and oestrogen, and the interpretation of these changes [94] was based on studies - outlined above - on LHRH release into portal blood.

In the bullfrog, LHRH or an LHRH-like substance is a transmitter in the lower lumbar sympathetic ganglia [49, 50].

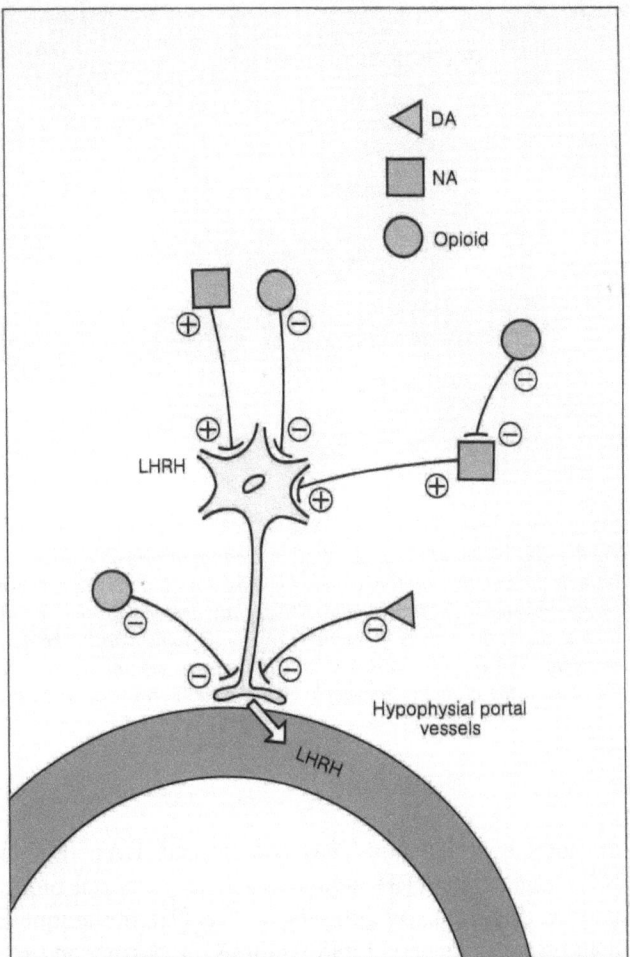

Fig. 7. How oestrogen could trigger the LHRH surge. It is unlikely that oestrogen has a direct affect on LHRH neurones, since the latter have no oestrogen receptors [94]. Oestrogen could stimulate noradrenergic (**NA**) neurones, which stimulate LHRH release and/or inhibit opioid, and dopaminergic (**DA**) neurones, which inhibit LHRH release. The + and − signs indicate stimulatory and inhibitory effects, respectively, of oestrogen at the cell bodies and of the neurones at synapses or points of contact

Thyrotropin-Releasing Hormone and Vasoactive Intestinal Peptide: Variable Actions and High Turnover

Thyrotropin-releasing hormone (TRH) mediates the neural control of thyrotropin release and when administered exogenously is a potent prolactin releasing factor (PRF), especially in man. Whether TRH is a physiological PRF remains uncertain. Immunoneutralization with anti-TRH serum does not block the prolactin surge induced by suckling in the rat, but it does delay significantly the onset of the spontaneous surge of prolactin on the afternoon of pro-oestrus [44, 91]. The latter finding, together with a significantly higher release of TRH into portal blood during the afternoon compared with the morning of pro-oestrus [33], suggests that TRH may play a physiological role in stimulating prolactin release on the afternoon of pro-oestrus. The neural systems that control prolactin are complex and multiple [57], and it is not surprising that TRH may possibly play a role under one condition but not another; this seems also to be the case for vasoactive intestinal peptide (VIP), which may play a crucial role in stimulating the prolactin surge in response to stress, but may not play a major role in the spontaneous pro-oestrous or suckling-induced surges of prolactin [1, 8].

There is a major difference between the release of TRH and that of somatostatin and LHRH in that, whereas the amount of somatostatin and LHRH released into portal blood per hour is much less than 1% of the total hypothalamic content, the amount of TRH released into portal blood per hour can be as much as 80% relative to

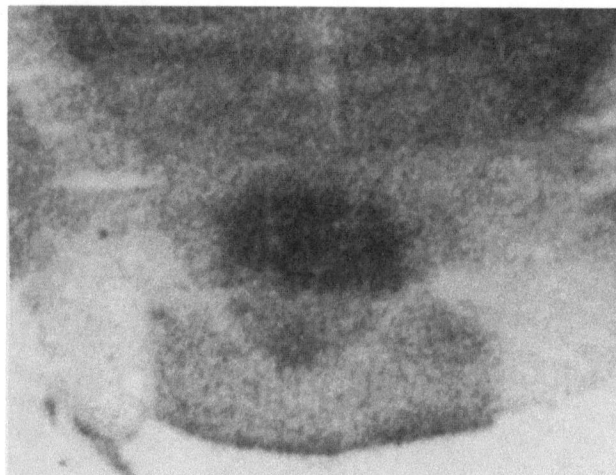 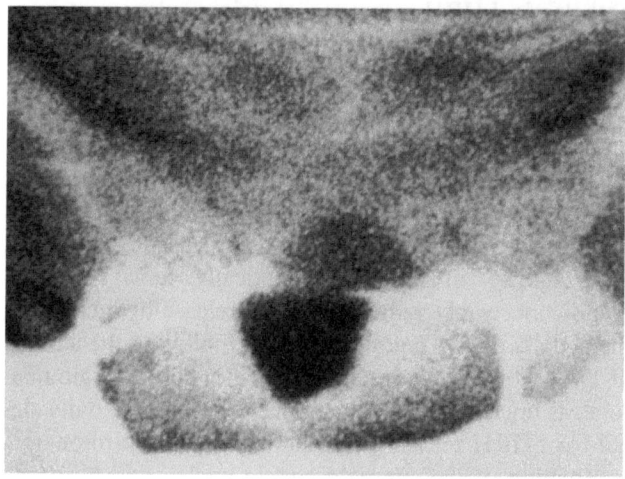

*Fig. 8. Application of [^{14}C]2-deoxyglucose autoradiography to determine metabolic activity in the central nervous system; autoradiograms of coronal sections through the pituitary gland of homozygous Brattleboro rats (which have hereditary diabetes insipidus). The section **on the left** is from a hydrated rat and shows the neural lobe (PN) slightly denser than the anterior lobe (AP). The section **on the right** is from an animal deprived of water for several hours: the activity of the nerve terminals in the neural lobe, as assessed by the much greater uptake of [^{14}C]2-deoxyglucose is increased about threefold. This can be reversed by administering a superactive agonist of vasopressin, which suggests that the increase in terminal activity is due to increased firing of the neurohypophysial neurones in response to an osmotic stimulus. MM, mamillary body. (From [101], with permission)*

the total hypothalamic TRH content [33]. Even though 50%–60% of the TRH immunoreactivity of portal blood may be due to factors other than TRH [93], the uniquely high turnover suggests rapid synthesis, processing and release of the tripeptide. One factor which may contribute to the high 'turnover' of TRH compared with LHRH and somatostatin is that both in frog skin [75] and in rat hypothalamus [39] the TRH precursor contains multiple copies of the tripeptide, while the precursors for somatostatin and LHRH contain only one copy per molecule of the active peptide.

Vasopressin and Oxytocin in the Neurohypophysis and in Hypophysial Portal Blood

Vasopressin and oxytocin are synthesized in the magnocellular neurones of the paraventricular and supraoptic nuclei, transported to the neurohypophysis and released into capillaries of the systemic circulation. The main function of vasopressin is the control of body water, while that of oxytocin is to stimulate milk ejection from the lactating breast and uterine contractions during parturition. The hypothalamo-neurohypophysial system was the first model used for studying the biosynthesis and release of neuropeptides [18]. The release of vasopressin and oxytocin is, as expected, triggered by action potentials that depolarize the nerve terminals: the frequency of the action potentials determines the amount of neuropeptide released [18]. The release of neurotransmitter involves the expenditure of considerable energy and this point can be shown by the fact that the uptake of [^{14}C]2-deoxyglucose, which is stoichometrically related to oxygen utilization, is significantly increased in the neurohypophysis by osmotic stimuli [100, 101] (Fig. 8). This increase in energy expenditure at the nerve terminals in the neurohypophysis correlates well with the firing rate of neurones that project to the neurohypophysis [21] but is not dependent upon the release of normal transmitter because a massive increase in [^{14}C]2-deoxyglucose uptake in the neurohypophysis also occurs in response to an osmotic challenge in homozygous Brattleboro rats [101] in which osmotic stimuli increase the firing rate of magnocellular neurones [22]. In the Brattleboro rat, a deletion of one nucleotide in the sequence of the neurophysin component of the vasopressin precursor results in a failure of the normal translation of the mRNA transcript or in packaging or processing of the precursor [85].

In addition to the neurohypophysis, vasopressin and oxytocin neurones project to other regions of the central nervous system, especially to the area of the brain stem thought to be involved in autonomic control and to the external layer of the ME. The vasopressin and oxytocin projections to the portal vessels are especially prominent in birds [68] and monotremes (G.C. Smith and G. Fink, unpublished data) but are also present in mammals [45]. The concentrations of vasopressin and oxytocin in hypophysial portal vessel blood of the rat and the rhesus monkey are much greater than those in the systemic circulation [45]. Vasopressin cannot be detected in portal blood collected from homozygous Brattleboro rats [45], and this is in keeping with the absence of vasopressin-containing terminals in the external layer of the ME in Brattleboro rats [97, 103]. These findings show that vasopressin-containing fibres of the ME, although distinct from those which project to the neurohypophysis, are nonetheless derived from neurones whose genomic mechanism for vasopressin synthesis is similar to that of the neurones which project to the neurohypophysis. It is not immediately obvious that this should be the case, because vasopressin-like material has been found in the ovaries [59] and adrenal glands [67] of the Brattleboro rat, suggesting that the defect in vasopressin may be tissue specific [7].

Removal of the whole pituitary gland does not alter the amount of vasopressin and oxytocin in portal blood, and the concentrations of these peptides in portal blood are far greater than can be explained by simple leakage from the cut ends of the hypothalamo-neurohypophysial fibres in the pituitary stalk. In fact, the concentrations of vasopressin and oxytocin in animals in which the supraoptico-hypophysial tract has been lesioned are significantly greater than the values in intact rats, probably because, as a result of the lesion (placed 7-10 days previously) there is hypertrophy of the neurohaemal junction at the ME and, as a consequence, an increase in the area of the synaptic contacts between the vasopressin and oxytocin neurones and the primary capillaries of the hypophysial portal vessels [45].

The role of oxytocin in hypophysial portal vessel blood remains to be established, but there is good evidence that vasopressin potentiates the action of corticotropin-releasing factor (CRF) and that vasopressin therefore plays a major role in the hypothalamic-pituitary-adrenal response to stress [37, 76].

Changes in Effector Cell Responsiveness

Studies on the pituitary gland have revealed another important principle which is generally applicable to the central nervous system, i.e. that the responsiveness of neuroeffector cells is not necessarily steady, but may alter significantly under different conditions.

Until the early 1970s the anterior pituitary gland was thought to operate as a steady system. However, studies with LHRH, TRH and CRF demonstrated that, quite to the contrary, the responsiveness of anterior pituitary cells to neurohormones changes quite dramatically under different physiological conditions [28]. Thus, for example, as already outlined above, the responsiveness of the pituitary gland to LHRH increases 20- to 50-fold before and during the spontaneous surge of LH [27, 28]. This increase in pituitary responsiveness to LHRH is initiated by the spontaneous, preovulatory surge of oestradiol-17β and is further increased by the priming effect of LHRH, the capacity of LHRH to increase the responsiveness of pituitary gonadotropes to itself. The priming effect of LHRH synchronizes the increasing concentrations of LHRH in portal blood with the increase in pituitary responsiveness, so that both events reach a peak at the same time and thus ensure the occurrence of a massive ovulatory surge of LH [27]. The priming effect of LHRH seems to be a unique property of this peptide, possibly because, apart from the oxytocin-uterine contraction system which operates dur-

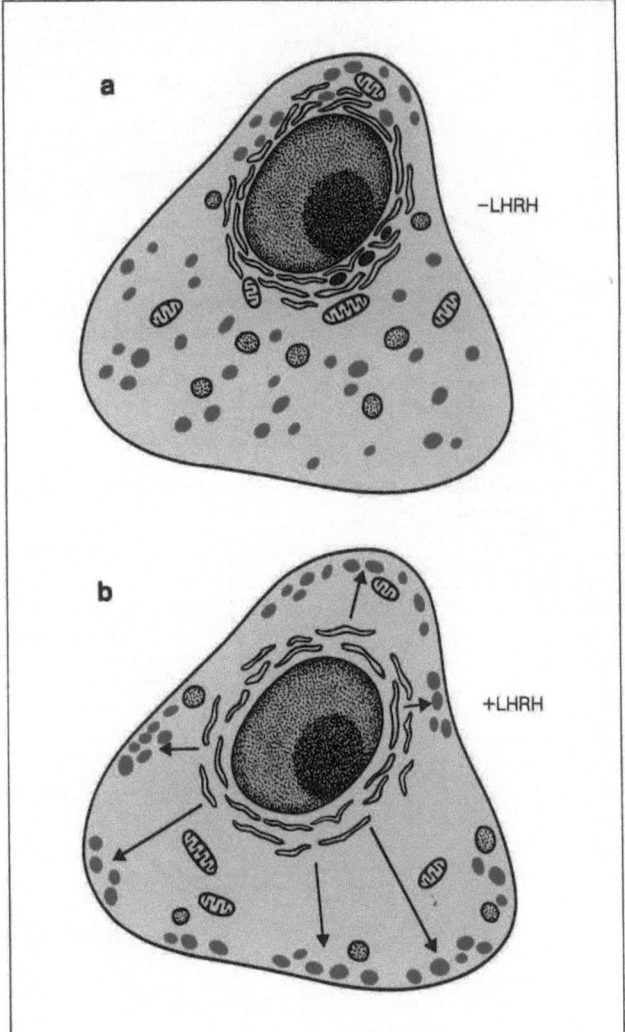

Fig. 9a, b. Sketches of pituitary gonadotropes as seen under the electron microscope. Gonadotrope a has not been exposed to LHRH: note that the secretory granules are scattered randomly throughout the cytoplasm. Gonadotrope b has been exposed to LHRH for 2 h: the granules are fewer in number (presumably due to release) and have moved close to the plasmalemma. The movement of granules to the plasmalemma, which is probably due to the activation of contractile elements of the cell, is the basis for the priming effect of LHRH. That is, LHRH stimulates margination of the secretory granules so that much more gonadotropin is available for release when the cells are exposed to another pulse of LHRH or to another secretagogue such as K^+ depolarization. (From [58])

ing parturition, the ovulatory surge of LH is the only positive-feedback endocrine system that operates under physiological conditions. The priming effect of LHRH can be demonstrated in vivo by different modes of administering exogenous LHRH and releasing endogenous LHRH [3, 30]. In vitro studies have revealed key differences between the releasing action and priming effect of LHRH. Thus, the releasing action but not the priming effect of LHRH can be mimicked by K^+ depolarization and Ca^{2+} ionophores, is dependent on normal concentrations of intra-

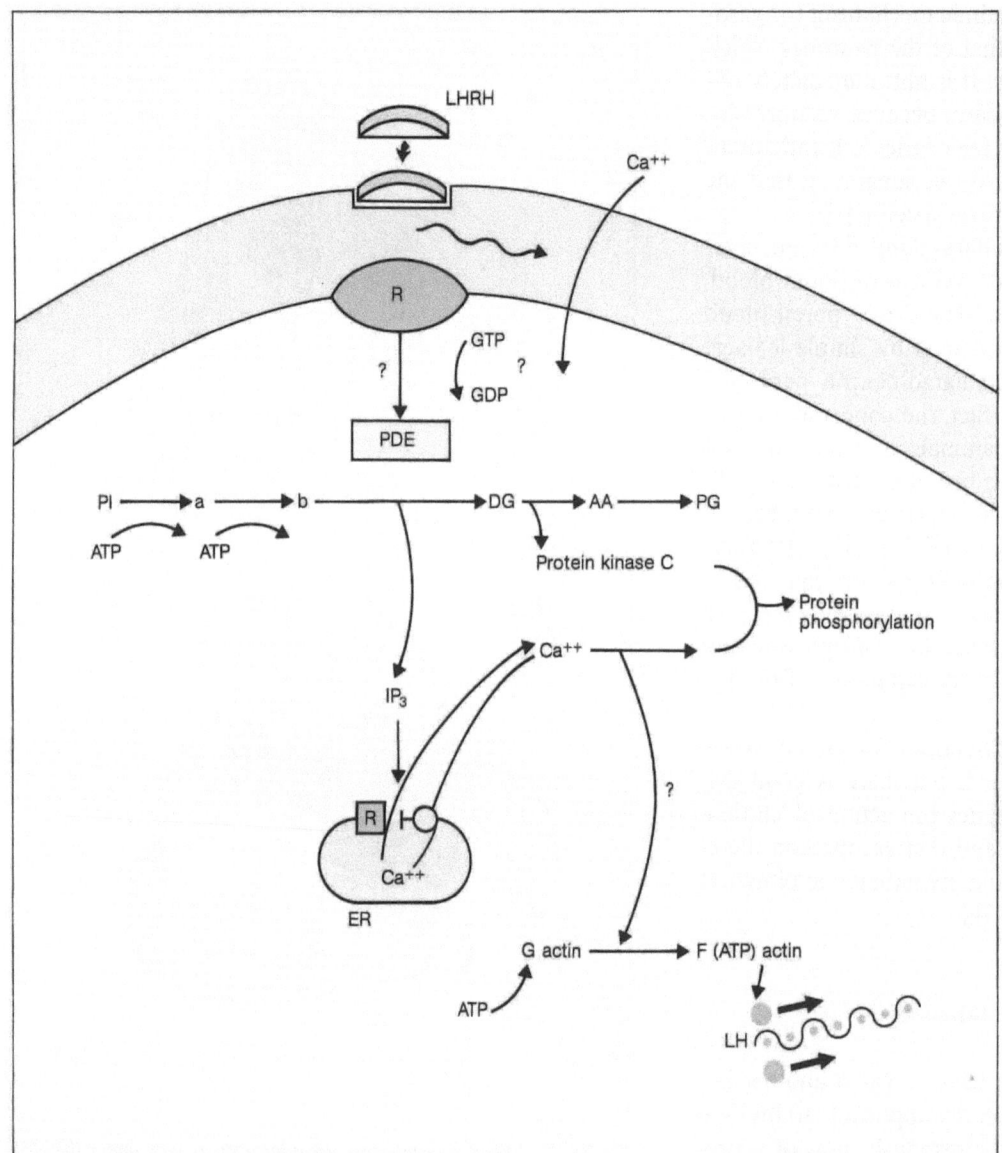

Fig. 10. Mechanism of action of LHRH. By interacting with its receptors, LHRH stimulates the influx of Ca^{2+} into the cell and triggers the hydrolysis of phosphatidylinositol (**PI**) by phosphodiesterase to form diacylglycerol (**DG**) and inositol triphosphate (**IP$_3$**), which further increases cytosolic Ca^{2+} concentrations by releasing Ca^{2+} from endoplasmic reticulum (**ER**) [6]. Increased cellular concentrations of Ca^{2+} interact with calmodulin and activate phospholipase A_2, leading to the formation of arachidonic acid (**AA**), and, as a consequence, the derivatives of AA, the prostaglandins (by the cyclooxygenase pathway) and the HETEs and leukotrienes (by the lipoxygenase pathway) [63]. Diacylglycerol activates protein kinase C, which is involved in protein phosphorylation and cell proliferation [6, 102] and which may also enhance the Ca^{2+} current [74]. Increased intracellular Ca^{2+} and activation of PI turnover may lead to activation of actin [10, 55]. Thus, it seems reasonable to propose that LHRH increases Ca^{2+} influx into the gonadotropes and increases the turnover of PI, which in turn leads to the release of Ca^{2+} from intracellular Ca^{2+} stores. This cascade involving phospholipid derivatives and Ca^{2+} brings about changes in the plasmalemma, the secretory granule membrane, and the cytoskeleton which lead to the immediate release of those gonadotrope granules which are close to the cell membrane - the releasing action of LHRH. The cascade, by way of facilitating the polymerization and activation of actin (involving the conversion of G to F actin), also brings more gonadotrope granules towards the cell membrane, so that subsequent exposure to a secretagogue leads to the release of much more gonadotropin - the priming effect of LHRH. Finally, the cascade is also likely to be involved in the stimulation by LHRH of gonadotropin synthesis. Although this scheme will no doubt prove to be an oversimplified view of how LHRH exerts its three major actions on gonadotropes, it does provide a basis for further detailed investigation.

How oestrogen and progesterone potentiate the gonadotropin response to LHRH also remains unclear, but these steroids could, by their genomic effects, increase the number of LHRH receptors and stimulate the production of proteins/enzymes involved in the packaging, intracellular transport and release of the gonadotropins, and in addition exert allosteric affects on enzyme action cellular Ca^{2+}, but is not inhibited by disruption of the microfilaments or blockade of protein synthesis. Recent studies have demonstrated that LHRH priming is associated with the de novo synthesis of a 69-K protein and with a shift in isoelectric points of two larger proteins, possibly due to their phosphorylation [20]. The priming effect probably involves the release of intracellular Ca^{2+} [20]. Ultrastructural studies have shown that the priming effect involves an increase in length and a change in the angle of the microfilaments in gonadotropes and a migration of secretory granules towards the plasmalemma of the gonadotrope [58]. This migration of granules leads to an increase in the pool of LH available for release, so that when the gonadotropes are exposed for a second time to a secretagogue such as K$^+$ depolarization, Ca^{2+} ionophores or LHRH itself, a massive second release of LH occurs [69, 70] (Fig. 9).

Second Messenger Systems for Peptide Neurotransmitters

Second messenger systems for neurotransmitter action have been discussed in detail on pp. 37ff. However, because LHRH has three major actions on the pituitary gonadotropes (stimulation of gonadotropin synthesis and release and the priming effect [27]), it may be instructive to consider whether the three actions could be mediated by the same second messengers.

Although cAMP may be involved in some of the late effects of LHRH [54], it does not act as second messenger for LHRH. Rather, the actions of LHRH are mediated by way of the Ca^{2+} and the phosphatidylinositol (PI) second messenger systems (Fig. 10).

Clinical Studies

With this background, I now wish to illustrate the way in which our unit has used the neuroendocrine window of the brain to investigate abnormalities in central neurotransmission in functional and organic psychoses. Our aim has been to examine systematically the spontaneous secretion of pituitary hormones in order to determine whether a single change or a constellation of changes in hormone secretion could identify defects in central neurotransmission that underlie the psychoses. These studies have been carried out on the basis of three principles designed to overcome some of the deficiencies of previous psychoneuroendocrine studies. First, the patients were all drug free for a period of at least 3 months before investigation. Second, changes in hormone secretion were compared between the psychoses in order to determine whether any particular change was selective or specific for any particular condition. Third, blood samples for

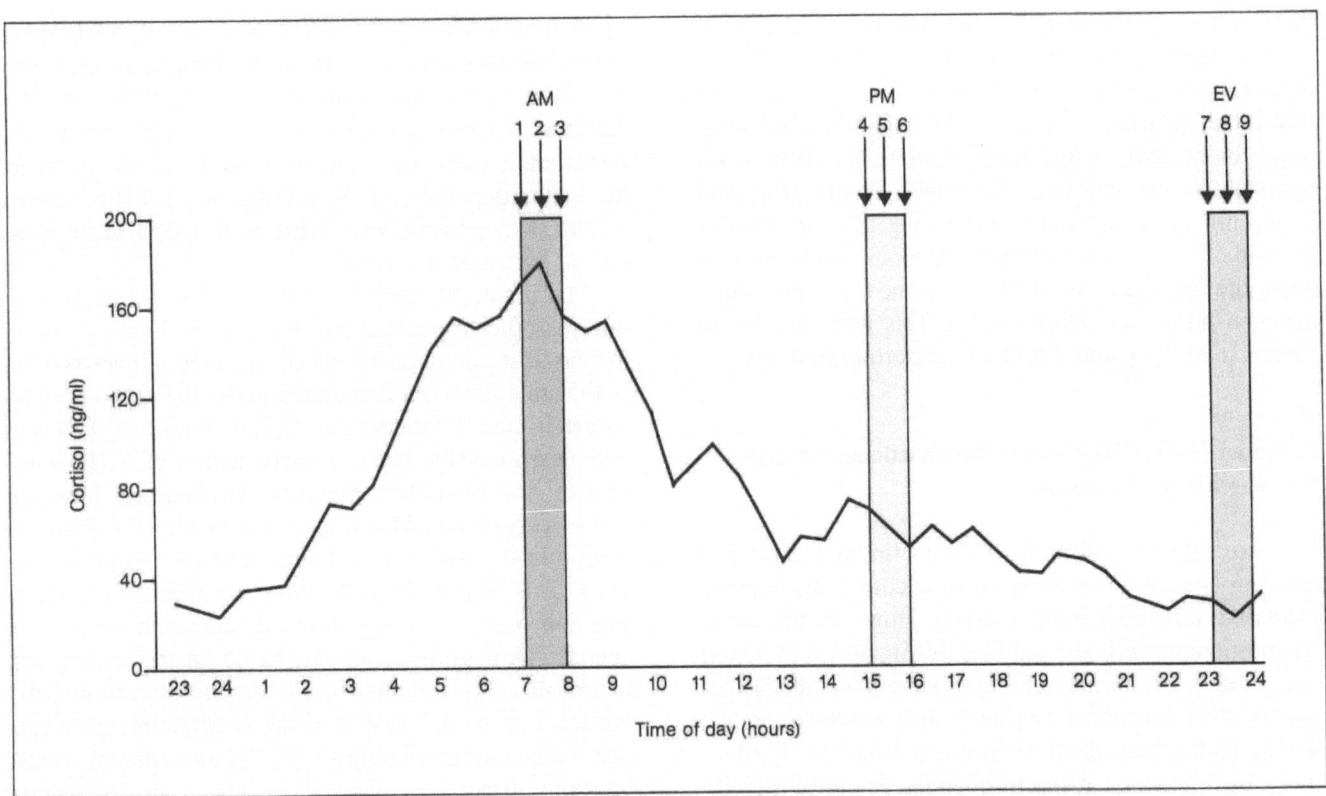

*Fig. 11. The 'neuroendocrine-day' sampling method. Blood samples are taken at 7:00, 7:30 and 8:00 o'clock (**AM**), at 3:00, 3:30 and 4:00 o'clock (**PM**), and at 11:00, 11:30 and 12:00 o'clock (**EV**). It can be seen from the mean plasma cortisol concentrations obtained from blood samples taken at 30-min intervals that the neuroendocrine-day sampling method takes into account the circadian patterns as well as fluctuations in hormone release*

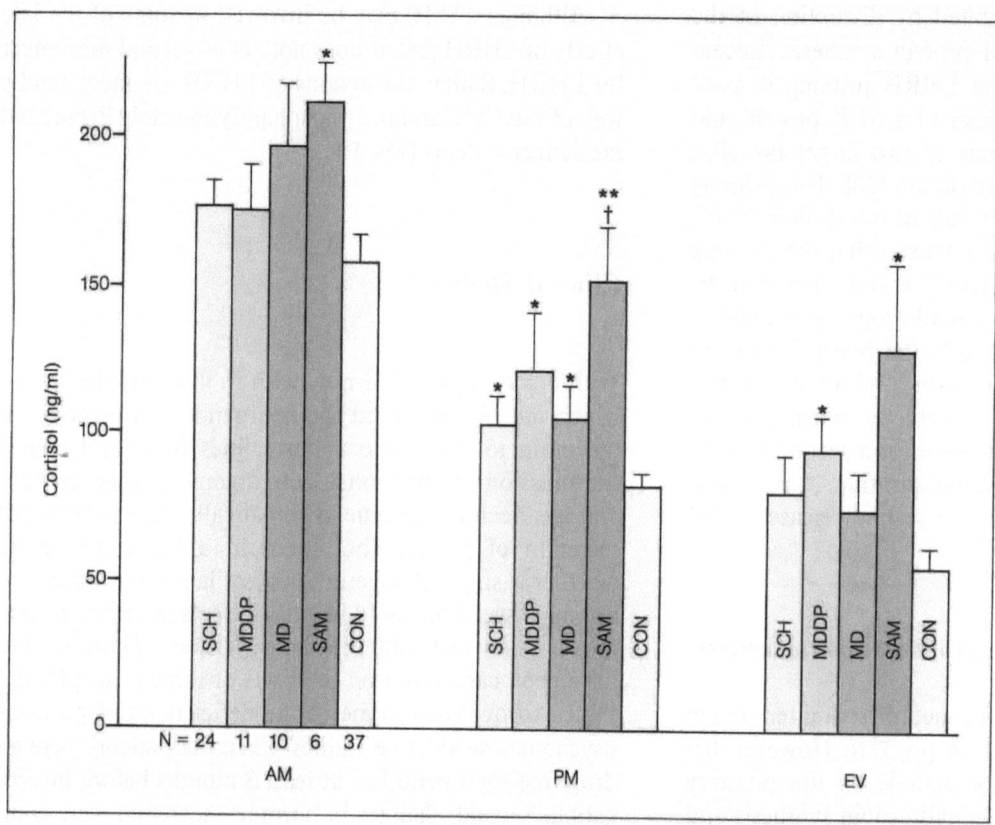

Fig. 12. Plasma cortisol concentrations in patients, classified according to the Research Diagnostic Criteria: schizophrenia (**SCH**), major depressive disorder psychotic subtype (**MDDP**), manic disorder (**MD**), schizoaffective disorder manic type (**SAM**), control subjects (**CON**). The means (± SEM) are calculated from the average of three values for each patient obtained during the three sampling periods of the 'neuroendocrine day' (see Fig. 11). Significant differences: *, $P < 0.05$, **, $P < 0.01$, compared with control subjects; +, $P < 0.05$ compared with schizophrenia. (From [15], with permission)

hormone estimation were taken at different times of the day to allow for both the circadian changes and fluctuations in hormone release. It is not possible to take blood samples from severely ill psychiatric patients at frequent intervals throughout a 24-h period. However, we found that it was possible to use a 'neuroendocrine-day' technique, by means of which blood samples are taken at 8-h intervals in the morning (7–8 a.m.), in the afternoon (3–4 p.m.) and at night (11–12 p.m (Fig. 11), and that this method of blood sampling would allow us to take into account the circadian pattern as well as the fluctuations that occur in hormone release [15, 16]. This work has led to several interesting and potentially important findings.

Increased Cortisol Secretion in the Functional Psychoses: Is it Specific for Depression?

We were able to confirm the previous findings of several other groups that in severe depression cortisol secretion is increased, especially in the evening. However, the cross-diagnostic approach showed that this increased secretion of cortisol is not specific for major depression but also occurs in other functional psychoses such as schizophrenia, mania and schizoaffective disorders (Fig. 12). Furthermore, by following patients through the course of their illness before therapy, during therapy and after recovery, we found that plasma concentrations of cortisol decrease with recovery from illness (Fig. 13). These two findings suggest that high plasma cortisol concentrations are a state-dependent feature of the functional psychoses and are not specific for the type of illness.

A question which has often been raised is why psychotic patients (mainly with major depressive disorder) who have hypercortisolaemia do not develop the physical features of Cushing's syndrome. Our studies show that peripheral (lymphocyte) glucocorticoid receptor numbers are low in depression [105], and this may explain, at least in part, why patients with depression appear to be resistant to hypercortisolaemia.

The increased cortisol secretion in manic depression and the other functional psychoses is likely to be due to increased secretion of ACTH consequent to increased hypothalamic drive. As mentioned under the section on vasopressin and oxytocin (above), the control of ACTH is complex, involving the synergistic action of CRH, vasopressin and possibly adrenaline. Psychiatrists have for many years accepted the dogma that increased cortisol secretion in depression was due to reduced central NA activity. This dogma fitted the oversimplifed view that depression was a consequence of underactivity of the central NA neurones, and was based on the finding that in the dog NA inhibited glucocorticoid secretion [104]. However, in man, infusion of the α-adrenoreceptor agonist methoxamine stimulated ACTH and cortisol release and this effect was blocked by the α-adrenoreceptor blocker, phentolamine [62]. The β-adrenoreceptor blocker, propranolol, had no effect on the action of methoxamine but did enhance the ACTH response to insulin-induced hypoglycaemia. The inference drawn from these results

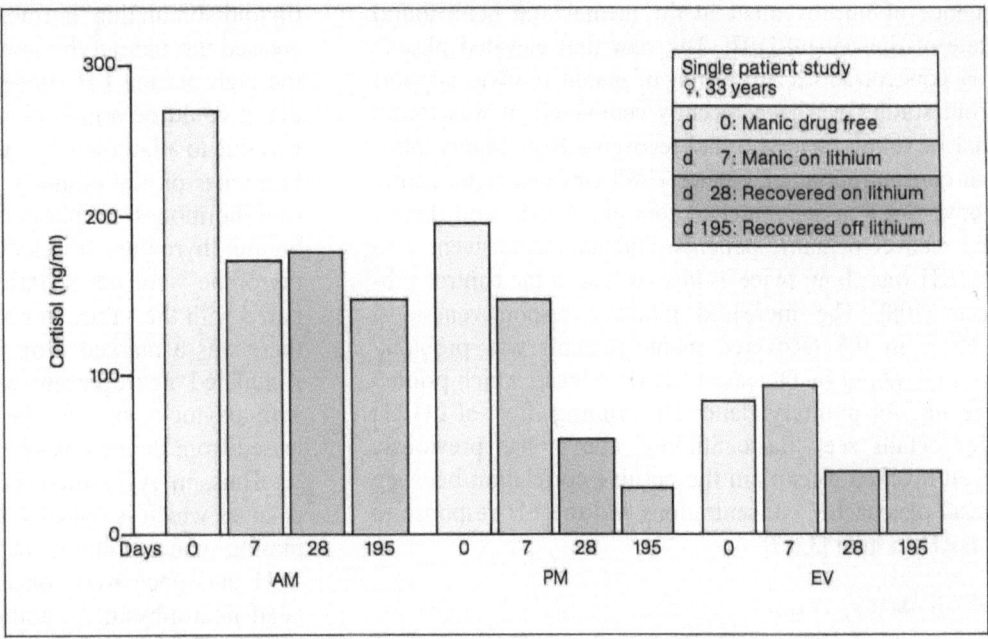

Fig. 13. Single-case study to illustrate the value of the neuroendocrine-day technique. The mean cortisol concentration of the nine blood samples taken during the 'neuroendocrine day' (see Fig. 11) are shown. The samples were taken from a female manic patient, and the figure shows that plasma cortisol concentrations decreased markedly on recovery

was that in man, an α-adrenoreceptor mechanism stimulates while a β-adrenoreceptor mechanism may inhibit ACTH and, as a consequence, cortisol release [62]. In the rat, as in man, NA stimulates corticosterone secretion apparently by a central action [2, 52].

High Plasma LH Concentrations in Young Men with Mania

A quite unexpected finding in our neuroendocrine survey of the major functional psychoses was that in young men with mania, plasma concentrations of LH were significantly higher than the plasma LH concentrations in age- and sex-matched control subjects [106] (Fig. 14). The plasma concentrations of LH in young men with schizophrenia were not different from those in control subjects. The increased plasma LH concentrations in the manic patients could not be attributed to a failure in negative feedback of androgens on the hypothalamic-pituitary system because the plasma concentrations of free and bound testosterone were not significantly different from the values in control subjects or in schizophrenics [106]. One particular case of a man who was in fact on drugs and had severe mania suggested to us that the high plasma LH concentrations in mania were *trait* rather than *state* dependent. The plasma concentrations of LH in this one case were six times greater on occasion than the highest values in age-matched control subjects and did not change either as a consequence of drug administration or as a conse-

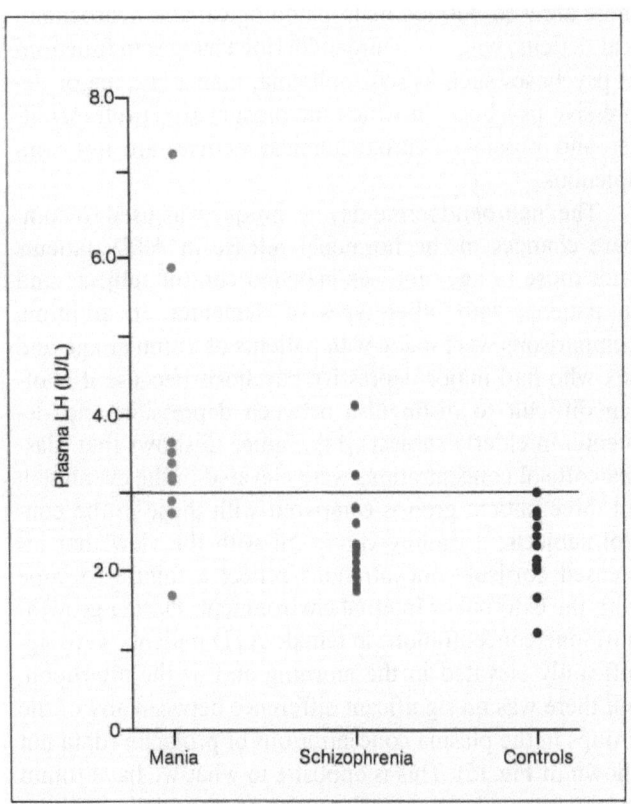

Fig. 14. Mean plasma luteinizing hormone (*LH*) concentrations based on nine estimations in 'neuroendocrine-day' studies carried out on young men with mania or schizophrenia or on normal controls. **Broken line**, Threshold based on highest value in controls. (From [106], with permission)

quence of improvement in the mental and behavioural state of the patient [107]. The view that elevated plasma LH concentrations are a trait of mania receives support from studies we have recently completed: it was found that in young men who had recovered from mania, plasma concentrations of LH were still elevated significantly above those in age-matched control subjects, and that in the recovered manic patients pituitary responsiveness to LHRH was about twice as high as that in the control subjects [108]. The increased pituitary responsiveness to LHRH in the recovered manic patients was probably brought about by increased LHRH release, which primed the anterior pituitary gland. The priming effect of LHRH (for details see "Basic Studies" above) has previously been invoked to explain the positive correlation between basal plasma LH concentrations and the LH response to LHRH in man [3, 77].

Neuroendocrine Changes in Alzheimer-type Dementia

Alzheimer-type Dementia (ATD) is characterized by macroscopic and microscopic changes in the brain and by a selective loss of central choline acetyltransferase, and therefore acetylcholine, noradrenaline, and somatostatin (see pp. 47ff. and Yates et al. [110]). For reasons that are obvious, it seemed that it would be exceedingly useful to compare the neuroendocrine changes in ATD, in which there are well-defined neuropathological and neurochemical deficits, with the neuroendocrine changes in functional psychoses such as schizophrenia, mania and major depressive psychosis in which the presumed neurotransmitter and possible neuroanatomical deficits are not conspicuous.

The 'neuroendocrine-day' technique was used to compare changes in the hormonal release in ATD patients with those in age- and sex-matched control subjects and in patients with other types of dementia. In addition, comparisons were made with patients of a similar age and sex who had major depressive psychosis because it is often difficult to distinguish between depression and dementia in elderly subjects [14]. Figure 15 shows that plasma cortisol concentrations were elevated in the evening in all three patient groups compared with those in the control subjects, a finding consistent with the view that increased cortisol concentrations reflect a failure to cope with the external or internal environment. Plasma growth-hormone concentrations in female ATD patients were significantly elevated in the morning and in the afternoon, but there was no significant difference between any of the groups in the plasma concentrations of prolactin (data not shown in Fig. 15). This is opposite to what we have found in patients with functional psychoses, especially in those with schizophrenia and schizoaffective mania, in whom plasma prolactin concentrations are elevated compared with the values in control subjects while growth-hormone concentrations remain unchanged. The most striking finding regarding ATD was that the plasma concentrations of thyroid-stimulating hormone (TSH) were markedly increased throughout the sampling period. As in the case of the high plasma LH concentrations in patients with mania, it could be argued that this increase in plasma TSH was due to a failure in negative feedback action of thyroid hormones on the pituitary gland. However, assay of thyroid hormones in plasma showed that both free and bound thyroxine, triiodothyronine and reverse triiodothyronine were not significantly different in ATD compared with the values in normal subjects. Finally, in ATD there was a marked drop in the secretion of oestrogen-stimulated neurophysin, a protein which is associated with oxytocin, one of the two main neuropeptides released from the neural lobe.

Thus, in ATD there is a constellation of hormonal changes which is selective for the disorder, i.e. increased plasma concentrations of growth hormone, increased TSH and decreased concentrations of oestrogen-stimulated neurophysin. In addition, there is the non-specific increase in plasma concentrations of cortisol in the evening, probably related to a failure to cope.

As well as being able to reveal state- and/or trait-dependent changes in global hormone secretion, the psychoneuroendocrine approach can also identify significant relationships between hormone changes and specific symptoms and signs. For example, in ATD there is an inverse correlation between the plasma concentrations of cortisol and the global psychological test score; higher plasma concentrations of cortisol are also associated with agnosia (inability to recognize objects) and loss of orientation. In other words, as the dementia becomes worse, plasma cortisol concentrations increase, and this emphasizes again that high plasma cortisol concentrations are associated with a failure to cope. An inverse correlation was also found between the plasma concentrations of growth hormone and the total global score for psychological testing, as well as with paired-associates learning and agnosia [14].

Neuroendocrine Changes in Response to Electroconvulsive Therapy

In addition to studies using the neuroendocrine-day technique, we have examined the neuroendocrine changes that follow electroconvulsive therapy (ECT). Although used for many years as an effective treatment for depression, the mechanism of ECT is not understood. On the basis of studies of the effects of electroconvulsive shock in animals it has been assumed that ECT enhances central monoaminergic activity [40]. As in the case of our neuroendocrine-day studies, our aim was to investigate whether a single change or a constellation of changes in hormone secretion in response to ECT could help to establish the mechanism of action of ECT. Our null hypothesis was that by causing a widespread seizure ECT would result in the massive release of all pituitary hormones, and we were surprised, therefore, that in the first

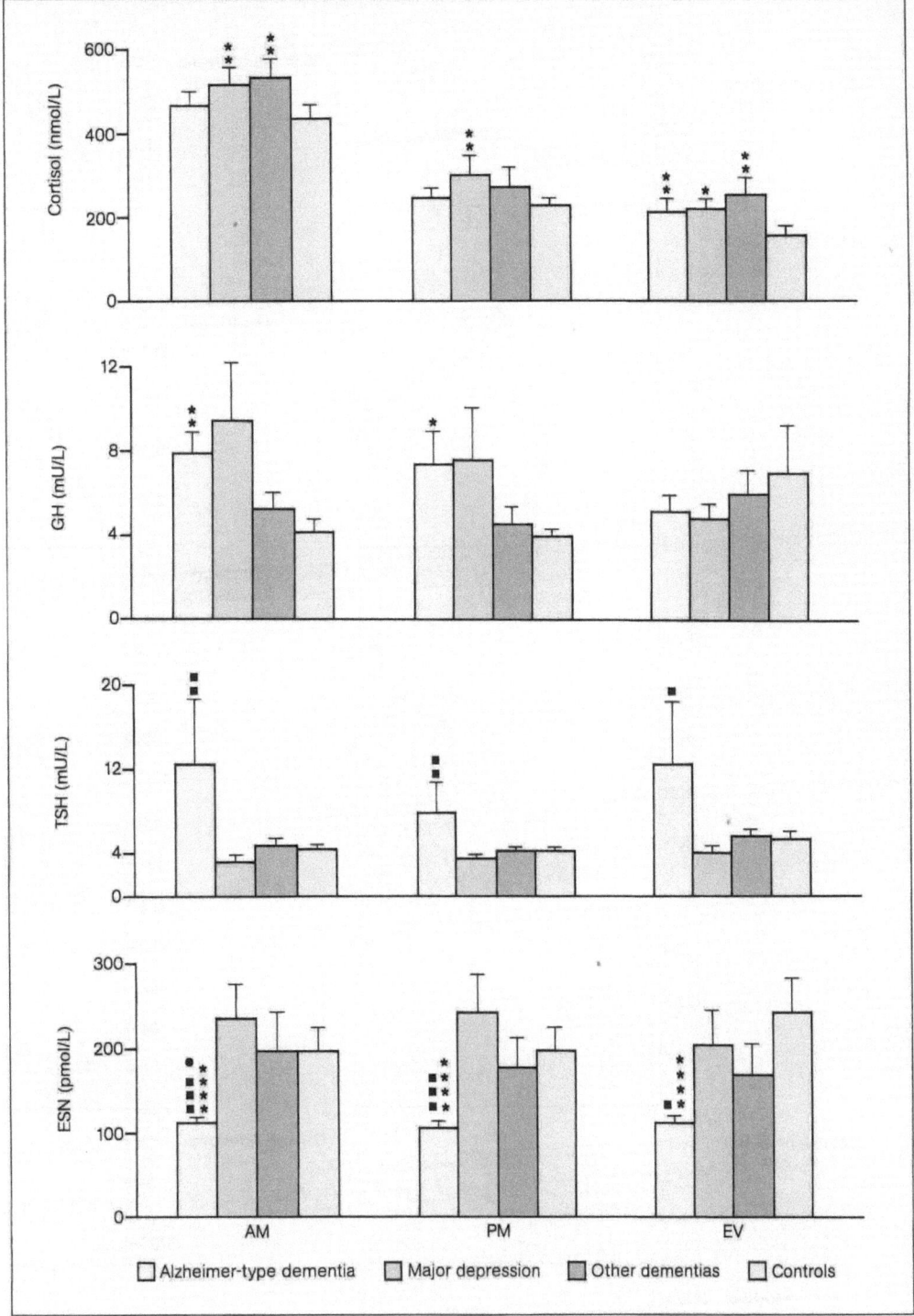

Fig. 15. The plasma concentrations of cortisol, growth hormone (**GH**), thyrotropin (**TSH**) and oestrogen-stimulated neurophysin (**ESN**) in patients with Alzheimer-type dementia (ATD; n = 17), major depression (n = 7-16), other dementias (n = 9), and control subjects (n = 11-37). Blood samples were collected by the 'neuroendocrine-day' technique (see Fig. 11) and the means of the three values in each of the three times were used to calculate the means (± SEM) for each group. Significant differences: patient values vs. control values; *, P< 0.05, **, P< 0.02, ***, P< 0.01, ****, P< 0.001; ATD vs. major depression; ■, P< 0.05, ■ ■, P< 0.02, ■ ■ ■, P< 0.01; ATD vs. other dementia; ●, P< 0.05. (Data from [14])

6 min after ECT there were selective increases only in prolactin and the neurophysins, and no change in the other so-called stress hormones: cortisol and growth hormone [109] (Fig. 16). Subsequent studies in which we have monitored hormone changes for 90 min have confirmed these findings, with the additional facts that (a) there is no significant change in the plasma concentrations of LH and TSH, (b) plasma growth-hormone concentrations decrease slightly after ECT, and (c) plasma ACTH increases immediately after ECT, followed about 15-30 min later by a significant increase in the plasma concentrations of cortisol.

These hormonal changes could not be attributed to the pharmacology or stress associated with the anaesthesia required for ECT since, except for an isolated increase in plasma prolactin concentrations, no significant hormonal changes occurred in mentally normal patients who were subjected to a similar anaesthetic and premedication

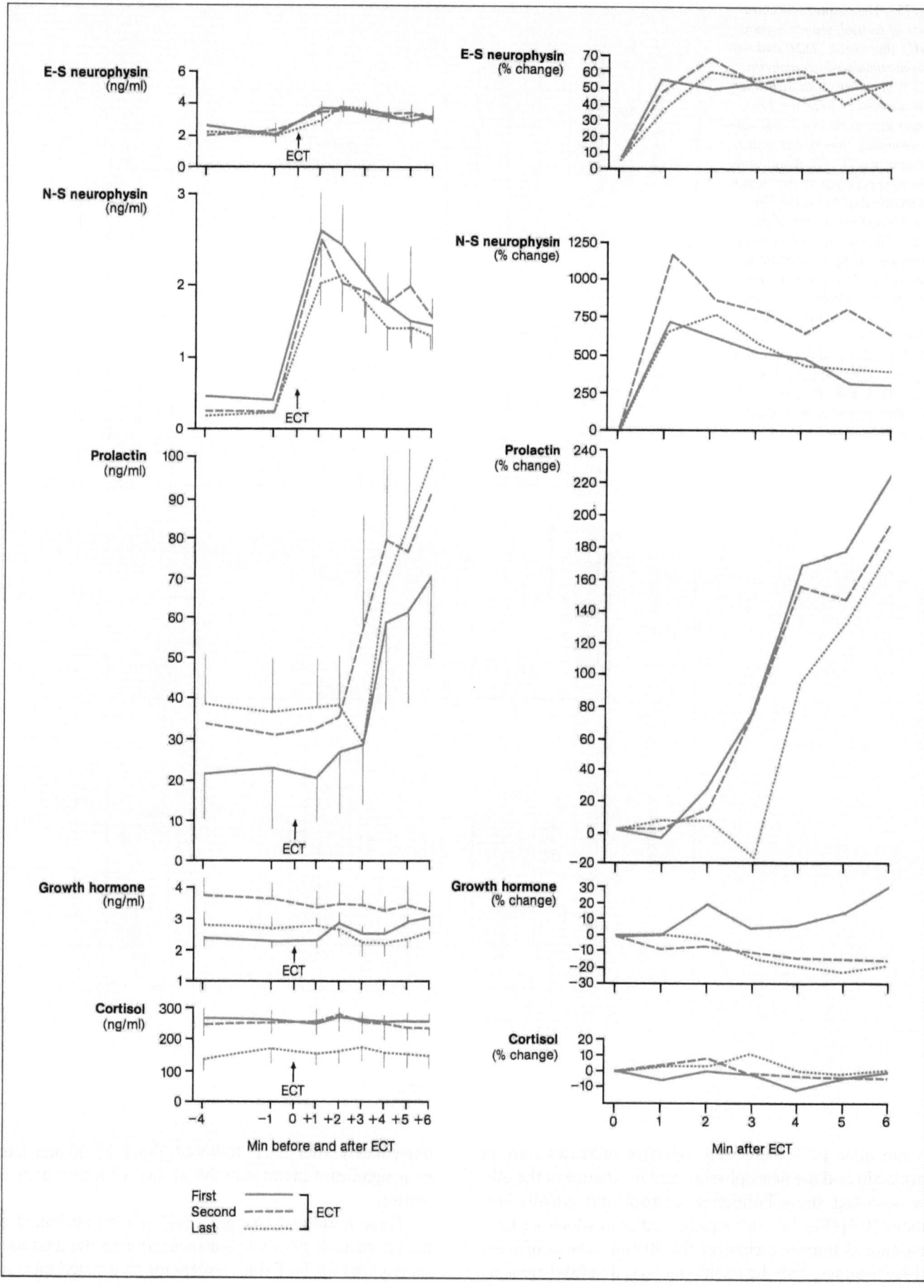

but were not given ECT. As might be expected, there were no significant changes in the peripheral plasma concentrations of LHRH or TRH after ECT [109].

Summary and Conclusions

This chapter has reviewed the way in which the hypothalamic-pituitary system can be used as a neuroendocrine window to investigate the characteristics of the release and action of neuropeptides with reference to LHRH, TRH, VIP, somatostatin, vasopressin and oxytocin. Although we are beginning to understand the physiology of these neuropeptides, the neuroendocrine window can be used in further extensive and intensive studies to answer such questions as: How does oestrogen exert its actions on NA and opioid neurones?; How are the surge and pulse generators for LHRH linked?; Why is the release of TRH relative to hypothalamic content so much greater than that of all the other known hypothalamic peptides?; Which of the several prolactin-releasing and -inhibiting factors predominate under different conditions?; and What is the functional significance of differential processing of the somatostatin precursor? As exemplified by LHRH, the characteristics of neurotransmitter release and action in the hypothalamic-pituitary system provide important clues as to how such or similar transmitters might be released and act in other parts of the nervous system.

The neuropeptides discussed in the present review have been reasonably well characterized, but a whole array of other neurohormones/neurotransmitters await further investigation. What, for example, is the role of the calcitonin gene-related peptide (CGRP) in the hypothalamus [4, 77]? Alternative processing of RNA transcripts from the calcitonin gene results in the production of a distinct mRNA encoding for CGRP in the mammalian hypothalamus and in an mRNA encoding for calcitonin in the thyroid gland. The precise mechanism of the alternative processing (for details see Chap. 4) and the question of why calcitonin appears to be present in large amounts in the ME/hypothalamus of reptiles and other submammalian species but not in mammals [60, 77] remain to be elucidated. However, as in the case of the 'established' neuropeptides reviewed here, CGRP was first discovered in the hypothalamus [4] and subsequently found to be widely distributed in mainly the sensory and autonomic components of the nervous system [77]. Also as in the case of the 'established' neuropeptides, our understanding of the CGRP/calcitonin gene system in the hypothalamus may provide vital clues as to how and why the system operates in other parts of the body.

What conclusions may be drawn from the results of the psychoneuroendocrine studies? Increased plasma cortisol concentrations would appear to be a state-dependent feature of all severe psychoses and are probably related to a failure to cope. The hypercortisolaemia is due to increased hypothalamic stimulation of ACTH release, but the mechanism of this increased hypothalamic activity, postulated to involve an NA system, has yet to be determined. The raised plasma concentrations of LH in patients with mania and the constellation of hormonal changes that occurs in patients with ATD may provide useful biochemical markers to assist in diagnosis. More importantly, however, these changes in hormone output reflect changes that occur in central neurotransmission in the two disorders. Thus, for example, the increased secretion of LH in mania may be due to overactivity of the NA neurones which drive the LHRH neurones, to underactivity of the DA and opioid neurones that inhibit LHRH release, or, possibly, to all three of these factors working in concert. A unitary hypothesis for the increased secretion of TSH and growth hormone in ATD might be that this is due to a decrease in the secretion of somatostatin (which inhibits TSH as well as growth-hormone release) into portal blood, consequent to a decrease in the brain content of somatostatin. With respect to the neuroendocrine changes after ECT, 'endocrine algebra' suggests that most of the main hormone effects seen could be explained by activation of central serotonergic neurones, a finding which is in agreement with the results of electroconvulsive shock in animals [40]. However, these explanations are greatly oversimplified and incomplete, and before a conclusion can be reached much more information on the neurotransmitter control of the hypothalamic neuroendocrine neurones is required. Indeed, if we understood precisely how the neuroendocrine neurones are controlled it would be possible, on the basis of our present neuroendocrine data, to explain, at least in large part, the neurochemistry of mania, ATD and ECT. Such information requires more experimental research into the mechanisms which regulate the synthesis, release and action of central neurotransmitters, and clinical pharmacological investigations which check systematically which of the neurotransmitter systems are at fault in causing abnormal neurohormone release into the hypophysial portal vessels and, as a consequence, pituitary hormone release into the peripheral circulation.

Acknowledgements. I am grateful to my colleagues, especially Drs. Lawrence J. Whalley, Janice E. Christie and Ivy M. Blackburn and Mr. John Bennie, Mr. Heinz Dick and Mrs. Graciela Sanchez Watts and the nursing staff of the Royal Edinburgh Hospital who were responsible for the clinical data reviewed in this chapter. My thanks go also to Miss Jo Donnelly for her careful preparation of the typescript.

Fig. 16. Mean (± SEM) plasma concentrations of oestrogen- and nicotine-stimulated neurophysin (*E-S neurophysin* and *N-S neurophysin* respectively), prolactin, growth hormone and cortisol before and immediately after the first, second and last ECT in eight women treated for depression. The mean percentage changes after ECT are also shown. (From [109], with permission)

References

1. Abe H, Engler D, Molitch ME, Bollinger-Gruber J, Reichlin S (1985) Vasoactive intestinal peptide is a physiological mediator of prolactin release in the rat. Endocrinology 116: 1383-1390
2. Abe K, Hiroshigi T (1974) Changes in plasma corticosterone and hypothalamic CRF levels following intraventricular injection or drug-induced changes of brain biogenic amines in the rat. Neuroendocrinology 14: 195-211
3. Aiyer MS, Fink G, Greig F (1974) Changes in sensitivity of the pituitary gland to luteinizing hormone releasing factor during the oestrous cycle of the rat. J Endocrinol 60: 47-64
4. Amara SG, Jonas V, Rosenfeld MG, Ong ES, Evans RM (1982) Alternative RNA processing in calcitonin gene expression generates mRNAs encoding different polypeptide products. Nature 298: 240-244
5. Barraclough CA, Wise PM (1982) The role of catecholamines in the regulation of pituitary luteinizing hormone and follicle-stimulating hormone secretion. Endocr Rev 3: 91-119
6. Berridge MJ, Irvine RF (1984) Inositol triphosphate, a novel second messenger in cellular signal transduction. Nature 312: 315-321
7. Bonner TI, Brownstein MJ (1984) Vasopressin, tissue-specific defects and the Brattleboro rat. Nature 310: 17
8. Brar AK, Fink G, Maletti M, Rostene W (1985) Vasoactive intestinal peptide in rat hypophysial portal blood: effects of electrical stimulation of various brain areas, the oestrous cycle and anaesthetics. J Endocrinol 106: 275-280
9. Brazeau P, Vale W, Burgus R, Ling N, Butcher M, Rivier J, Guillemin R (1973) Hypothalamic polypeptide that inhibits the secretion of immunoreactive pituitary growth hormone. Science 179: 77-79
10. Burn P, Rotman A, Meyer RK, Burger MM (1985) Diacylglycerol in large α-actinin/actin complexes and in the cytoskeleton of activated platelets. Nature 314: 469-472
11. Carmel PW, Araki S, Ferin M (1976) Pituitary stalk portal blood collection in rhesus monkeys: evidence for pulsatile release of gonadotropin-releasing hormone (GnRH). Endocrinology 99: 243-248
12. Chiappa SA, Fink G, Sherwood NM (1977) Immunoreactive luteinizing hormone releasing factor (LRF) in pituitary stalk plasma from female rats: effects of stimulating diencephalon, hippocampus and amygdala. J Physiol 267: 625-640
13. Ching M (1982) Correlative surges of LHRH, LH and FSH in pituitary stalk plasma and systemic plasma of rat during proestrus. Neuroendocrinology 34: 279-285
14. Christie JE, Whalley LJ, Bennie J, Dick H, Blackburn IM, Blackwood DHR, Fink G (1987) Characteristic plasma hormone changes in Alzheimer's disease. Br J Psychiatry 150: 674-681
15. Christie JE, Whalley LJ, Dick H, Blackwood DHR, Blackburn IM, Fink G (1986) Raised plasma cortisol concentrations are a feature of drug-free psychotics and are not specific for depression. Br J Psychiatry 148: 58-65
16. Christie JE, Whalley LJ, Dick H, Fink G (1983) Plasma cortisol concentrations in the functional psychoses and Alzheimer-type dementia: a neuroendocrine-day approach in drug-free patients. J Steroid Biochem 19: 247-250
17. Clarke IJ, Cummins JT (1982) The temporal relationship between gonadotropin releasing hormone (GnRH) and luteinizing hormone (LH) secretion in ovariectomized ewes. Endocrinology 111: 1737-1739
18. Cross BA, Dyball REJ, Dyer RG, Jones CW, Lincoln DW, Morris JF, Pickering BT (1975) Endocrine neurons. Recent Prog Horm Res 31: 243-286
19. Curtis A, Lyons V, Fink G (1983) The human hypothalamic LHRH precursor is the same size as that in rat and mouse hypothalamus. Biochem Biophys Res Commun 117: 872-877
20. Curtis A, Lyons V, Fink G (1985) The priming effect of LH-releasing hormone: effects of cold and involvement of new protein synthesis. J Endocrinol 105: 163-168
21. Dyball REJ (1974) Single unit activity in the hypothalamo-neurohypophysial system of the Brattleboro rat. J Endocrinol 60: 135-143
22. Dyball REJ, Leng G (1982) Osmoresponsive cells in the supraoptic nucleus of the Brattleboro rat. J Physiol (Lond) 332: 89P-90P
23. Dyer RG, Mansfield S, Yates JO (1980) Discharge of gonadotrophin-releasing hormone from the mediobasal part of the hypothalamus: effect of stimulation frequency and gonadal steroids. Exp Brain Res 39: 453-460
24. Eskay RL, Mical RS, Porter JC (1977) Relationship between luteinizing hormone releasing hormone concentration in hypophysial portal blood and luteinizing hormone release in intact, castrated and electrochemically stimulated rats. Endocrinology 100: 263-270
25. Everett JW, Sawyer CH (1950) A 24-hour periodicity in the 'LH-release apparatus' of female rats, disclosed by barbiturate sedation. Endocrinology 47: 198-218
26. Fink G (1976) The development of the releasing factor concept. Clin Endocrinol (Oxf) 5: [Suppl] 245-260
27. Fink G (1979) Neuroendocrine control of gonadotrophin secretion. Br Med Bull 35: 155-160
28. Fink G (1979) Feedback actions of target hormones on hypothalamus and pituitary with special reference to gonadal steroids. Ann Rev Physiol 41: 571-585
29. Fink G, Aiyer M, Chiappa S, Henderson S, Jamieson M, Levy-Perez V, Pickering A, Sarkar D, Sherwood N, Speight A, Watts A (1982) Gonadotropin-releasing hormone: release into hypophyseal portal blood and mechanism of action. In: McKerns K, Pantic V (eds) Hormonally active brain peptides: structure and function. Plenum, New York, pp 397-426
30. Fink G, Chiappa SA, Aiyer MS (1976) Priming effect of luteinizing hormone release factor elicited by preoptic stimulation and by intravenous infusion and multiple injections of the synthetic decapeptide. J Endocrinol 69: 359-372
31. Fink G, Geffen LB (1978) The hypothalamo-hypophysial system: model for central peptidergic and monoaminergic transmission. In: Porter R (ed) International review of physiology, neurophysiology III, vol 17. University Park Press, Baltimore, pp 1-48
32. Fink G, Jamieson MG (1976) Immunoreactive luteinizing hormone releasing factor in rat pituitary stalk blood: effects of electrical stimulation of the medial preoptic area. J Endocrinol 68: 71-87
33. Fink G, Koch Y, Ben-Aroya N (1982) Release of thyrotropin-releasing hormone into hypophysial portal blood is high relative to other neuropeptides and may be related to prolactin secretion. Brain Res 243: 186-189
34. Fink G, Smith GC (1971) Ultrastructural features of the developing hypothalamo-hypophysial axis in the rat: a correlative study. Z Zellforsch 119: 208-226
35. Fox SR, Smith MS (1985) Changes in the pulsatile pattern of luteinizing hormone secretion during the rat estrous cycle. Endocrinology 116: 1485-1492
36. Fuxe K, Hökfelt T (1970) Participation of central monoamine neurons in the regulation of anterior pituitary function with special regard to the neuro-endocrine role of tuberoinfundibular dopamine neurons. In: Bargmann W, Sharrer B (eds) Aspects of neuroendocrinology. Springer, Berlin Heidelberg New York, pp 192-205
37. Gillies GE, Linton EA, Lowry PJ (1982) Corticotropin-releasing activity of the new CRF is potentiated several times by vasopressin. Nature 299: 355-357
38. Goodman RH, Aron DC, Roos BA (1983) Rat pre-prosomatostatin. J Biol Chem 258: 5570-5573
39. Goodman RH, Monminy MR, Low MJ, Tsukada T, Fink S, Lechan RM, Wu P, Jackson IMD, Mandel G (1986) Biosynthesis

of somatostatin, vasoactive intestinal polypeptide, and thyrotropin-releasing hormone. In: Fink G, Harmar AJ, McKerns KW (eds) Neuroendocrine molecular biology. Plenum, New York pp 159-173
40. Grahame-Smith DG, Green R, Costain DW (1978) Mechanism of the anti-depressant action of electroconvulsive therapy. Lancet I: 254-256
41. Harmar AJ, Pierotti AR (1984) The pattern of molecular forms of somatostatin released by the rat median eminence differs from that released by the hypothalamus as a whole. J Physiol 357: 95 P
42. Harris GW (1955) Neural control of the pituitary gland. Edward Arnold, London
43. Horn AM, Fink G (1985) Parachlorophenylalanine blocks the spontaneous pro-oestrous surge of prolactin as well as LH and affects the secretion of oestradiol-17β. J Endocrinol 104: 415-418
44. Horn AM, Fraser HM, Fink G (1985) Effects of antiserum to thyrotrophin-releasing hormone on the concentrations of plasma prolactin, thyrotrophin and LH in the pro-oestrous rat. J Endocrinol 104: 205-209
45. Horn AM, Robinson ICAF, Fink G (1985) Oxytocin and vasopressin in rat hypophysial portal blood: experimental studies in normal and Brattleboro rats. J Endocrinol 104: 211-224
46. Jackson GL (1972) Effect of actinomycin D on estrogen-induced release of luteinizing hormone in ovariectomized rats. Endocrinology 91: 1284-1287
47. Jackson GL (1973) Time interval between injection of estradiol benzoate and LH release in the rat and effect of actinomycin D or cycloheximide. Endocrinology 93: 887-892
48. Jamieson MG, Fink G (1976) Parameters of electrical stimulation of the medial preoptic area for release of gonadotrophins in male rats. J Endocrinol 68: 57-70
49. Jan YN, Jan LY (1983) Electrophysiological techniques. In: Krieger DT, Brownstein MJ, Martin JB (eds) Brain peptides. Wiley, New York, pp 547-563
50. Jan YN, Jan LY, Kuffler SW (1980) A peptide as a possible transmitter in sympathetic ganglia of the frog. Proc Natl Acad Sci USA 76: 1501-1505
51. Kalra SP, Kalra PS (1983) Neural regulation of luteinizing hormone secretion in the rat. Endocr Rev 4: 311-351
52. Kawa A, Taniguchi Y, Mizuguchi I, Ryu S, Ariyama T, Kamisaki T, Koreeda F, Kanehisa T (1978) Effect of intraventricular administration of noradrenaline and dopamine on the levels of corticosterone in rats and denervation hypersensitivity resulting from intraventricular administration of 6-hydroxydopamine. Life Sci 23: 991-998
53. Knobil E (1980) The neuroendocrine control of the menstrual cycle. Recent Prog Horm Res 36: 53-58
54. de Koning J, van Dieten AMJ, Tijssen AMI, van Rees GP (1981) Dependence on protein synthesis of the N^6-monobutyryl cylcic AMP plus thophylline-mediated release of luteinizing hormone induced by luteinizing hormone releasing hormone from rat pituitary glands in vitro. J Endocrinol 88: 329-338
55. Lassing I, Lindberg U (1985) Specific interaction between phosphatidyl-inositol 4,5-biphosphate and profilactin. Nature 314: 472-474
56. Legan SJ, Karsh FJ (1975) A daily signal for the LH surge in the rat. Endocrinology 96: 57-62
57. Leong DA, Frawley LS. Neill JD (1983) Neuroendocrine control of prolactin secretion. Annu Rev Physiol 45: 109-127
58. Lewis CE, Morris JF, Fink G (1985) The role of microfilaments in the priming effect of LH-releasing hormone: an ultrastructural study using cytochalasin B. J Endocrinol 106: 211-218
59. Lim ATW, Lolait SJ, Barlow JW, Autelitano DJ, Toh BH, Boublik J, Abraham J, Johnston CI, Funder JW (1984) Immunoreactive arginine-vasopressin in Brattleboro rat ovary. Nature 310: 61-64
60. MacInnes DG, Laszlo I, MacIntyre I, Fink G (1982) Salmon calcitonin in lizard brain: a possible neuroendocrine transmitter. Brain Res 251: 371-373
61. Millar RP, Sheward WJ, Wegener I, Fink G (1983) Somatostatin-28 is a hormonally active peptide released into hypophysial portal vessel blood. Brain Res 260: 334-337
62. Nakai Y, Imura H, Yoshimi T, Matsukura S (1973) Adrenergic control mechanism for ACTH secretion in man. Acta Endocrinol (Copenh) 74: 263-270
63. Naor Z, Amsterdam A, Catt KJ (1984) Binding and activation of gonadotropin-releasing hormone receptors in pituitary gonadotrophs. In: Saxena BB, Birnbaumber KJ, Lutz L, Martini L (eds) Hormone receptors in growth and reproduction. Raven, New York, pp 1-19
64. Neill JD, Dailey RA, Tsou RC, Patton J, Tindall G (1976) In: Crosignani PG, Mischell DR (eds) Ovulation in the human. Academic, New York, pp 115-125
65. Neill JD, Patton JM, Dailey RA, Tsou RC, Tindall GT (1977) Luteinizing hormone releasing hormone (LHRH) in pituitary stalk blood of rhesus monkeys: relationship to level of LH release. Endocrinology 101: 430-434
66. Nikolics K, Mason AJ, Szonyi E, Ramachandran J, Seeburg PH (1985) A prolactin-inhibiting factor within the precursor for human gonadotropin-releasing hormone. Nature 316: 511-517
67. Nussey SS, Ang VTY, Jenkins JS, Choudrey HS, Bisset GW (1984) Brattleboro rat adrenal contains vasopressin. Nature 310: 64-66
68. Oksche A, Kirschstein H, Hartwig HG, Oehmke HJ, Farner DS (1974) Secretory parvocellular neurons in the rostral hypothalamus and in the tuberal complex of *Passer domesticus*. Cell Tissue Res 149: 363-369
69. Pickering AJMC, Fink G (1979) Priming effect of luteinizing hormone releasing factor; role of protein synthesis, contractile elements, Ca^{2+} and cyclic AMP. J Endocrinol 81: 223-234
70. Pickering AJMC, Fink G (1979) Variation in size of the 'readily releasable pool' of luteinizing hormone during the oestrous cycle of the rat. J Endocrinol 83: 53-59
71. Pierotti AR, Harmar AJ (1985) Multiple forms of somatostatin-like immunoreactivity in the hypothalamus and amygdala of the rat: selective localization of somatostatin-28 in the median eminence. J Endocrinol 105: 383-389
72. Plant TM, Dubey AK (1984) Evidence from the rhesus monkey *(Macaca mulatta)* for the view that negative-feedback control of luteinizing hormone secretion by the testis is mediated by a deceleration of hypothalamic gonadotropin-releasing hormone pulse frequency. Endocrinology 115: 2145-2153
73. Pradayrol L, Jornvall H, Mutt V, Ribet A (1980) N-terminally extended somatostatin: the primary structure of somatostatin-28. FEBS Lett 109: 55-58
74. De Reimer SA, Strong JA, Albert KA, Greengard P, Kaczmarek LK (1985) Enhancement of calcium current in *Aplysia* neurones by phorbol ester and protein kinase C. Nature 313: 313-316
75. Richter K, Kawashima E, Egger R, Kreil G (1984) Biosynthesis of thyrotropin-releasing hormone in the skin of *Xenopus laevis*: partial sequence of the precursor deduced from cloned cDNA. EMBO J 3: 617-621
76. Rivier C, Vale W (1983) Interaction of corticotropin-releasing factor and arginine vasopressin on adrenocorticotropin secretion in vivo. Endocrinology 113: 939-942
77. Rosenfeld MG, Mermod J-J, Amara SG, Swanson LW, Sawchenko PE, Rivier J, Vale WW, Evans RM (1983) Production of a novel neuropeptide encoded by the calcitonin gene via tissue-specific RNA processing. Nature 304: 129-135
78. Roth JC, Kelch RP, Kaplan SL, Grumbach MM (1972) FSH and LH response to luteinizing hormone releasing factor in prepubertal and pubertal children, adult males and patients with hypogonadotropic and hypergonadotropic hypogonadism. J Clin Endocrinol Metab 35: 926-930
79. Sakuma Y, Pfaff DW (1983) Modulation of the lordosis reflex of female rats by LHRH, its antiserum and analogs in the mesencephalic central gray. Neuroendocrinology 36: 218-224

80. Sarkar DK, Chiappa SA, Fink G, Sherwood NM (1976) Gonadotropin-releasing hormone surge in pro-oestrous rats. Nature 264: 461-463
81. Sarkar DK, Fink G (1979) Effects of gonadal steroids on output of luteinizing hormone releasing factor into pituitary stalk blood in the female rat. J Endocrinol 80: 303-313
82. Sarkar DK, Fink G (1979) Mechanism of the first spontaneous gonadotropin surge and that induced by pregnant mare serum and effects of neonatal androgen. J Endocrinol 83: 339-354
83. Sarkar DK, Fink G (1980) Luteinizing hormone releasing factor in pituitary stalk plasma from long-term ovariectomized rats: effects of steroids. J Endocrinol 86: 511-524
84. Sarkar DK, Fink G (1981) Gonadotropin-releasing hormone surge: possible modulation through postsynaptic α-adrenoreceptors and two pharmacologically distinct dopamine receptors. Endocrinology 108: 862-867
85. Schmale H, Richter D (1984) Single base deletion in the vasopressin gene is the cause of diabetes insipidus in Brattleboro rats. Nature 308: 705-709
86. Schuiling GA, Pols-Valhor N, van der Schaof-Verdonk GCJ, Koiten TR (1984) Blockade of LH and FSH secretion by LH-releasing hormone, by the LH-releasing hormone analogue, buserelin, and by combined treatment with LH-releasing hormone and oestradiol benzoate. J Endocrinol 103: 301-309
87. Seeburg PH, Adelman JP (1984) Characterization of cDNA for precursor of human luteinizing hormone releasing hormone. Nature 311: 666-668
88. Shen L-P, Pictet RL, Rutter WJ (1982) Human somatostatin. I. sequence of the cDNA. Proc Natl Acad Sci USA 79: 4575-4579
89. Sherwood NM, Chiappa SA, Sarkar DK, Fink G (1980) Gonadotropin-releasing hormone (GnRH) in pituitary stalk blood from proestrous rats: effects of anesthetics and relationship between stored and released GnRH and luteinizing hormone. Endocrinology 107: 1410-1417
90. Sheward WJ, Benoit R, Fink G (1984) Somatostatin-28(1-12)-like immunoreactive substance is secreted into hypophysial portal vessel blood in the rat. Neuroendocrinology 38: 88-90
91. Sheward WJ, Fraser HM, Fink G (1985) Effect of immunoneutralization of thyrotrophin-releasing hormone on the release of thyrotrophin and prolactin during suckling or in response to electrical stimulation of the hypothalamus in the anaesthetized rat. J Endocrinol 106: 113-119
92. Sheward WJ, Harmar AJ, Fink G (1985) LH-RH in the rat and mouse hypothalamus and rat hypophysial portal blood: confirmation of identity by high-performance liquid chromatography. Brain Res 345: 362-365
93. Sheward WJ, Harmar AJ, Fraser HM, Fink G (1983) Thyrotropin-releasing hormone in rat pituitary stalk blood and hypothalamus: studies with high-performance liquid chromatography. Endocrinology 113: 1865-1869
94. Shivers BD, Harlan RE, Morell JI, Pfaff DW (1983) Immunocytochemical localization of luteinizing hormone-releasing hormone in male and female rat brains. Neuroendocrinology 36: 1-12
95. Shivers BD, Harlan RE, Pfaff DW (1983) Reproduction: the central nervous system role of luteinizing hormone releasing hormone. In: Krieger DT, Brownstein MJ, Martin JB (eds) Brain peptides. Wiley, New York, pp 389-412
96. Sirinathsinghji DJS, Whittington PE, Audsley A, Fraser HM (1983) β-Endorphin regulates lordosis in female rats by modulating LH-RH release. Nature 301: 62-64
97. Sokol HW, Zimmerman EA, Sawyer WH, Robinson AG (1976) The hypothalamic-neurohypophysial system of the rat: localization and quantitation of neurophysin by light-microscopic immunocytochemistry in normal rats and in Brattleboro rats deficient in vasopressin and a neurophysin. Endocrinology 98: 1176-1188
98. Speight A, Popkin R, Watts AG, Fink G (1981) Oestradiol-17β increases pituitary responsiveness by a mechanism that involves the release and the priming effect of luteinizing hormone releasing factor. J Endocrinol 88: 301-308
99. Sutherland RC, Fink G, Charlton HM (1984) Effect of mating on the metabolic activity of the brain and pituitary gland assessed by [^{14}C]2-deoxyglucose in a reflex ovulator, the vole (Microtus agrestis). Brain Res 311: 317-322
100. Sutherland RC, Fink G, Morris JF (1985) Effects of water deprivation and deamino [8-D-arginine] vasopressin on [^{14}C]2-deoxyglucose uptake by the hypothalamo-hypophysial system in mice with hereditary nephrogenic diabetes insipidus. Brain Res 340: 297-303
101. Sutherland RC, Martin MJ, McQueen JK, Fink G (1983) Water deprivation results in increased 2-deoxyglucose uptake by paraventricular neurones as well as pars nervosa in Wistar and Brattleboro rats. Brain Res 271: 101-108
102. Truneh A, Albert F, Golstein P, Schmitt-Verhulst A-M (1985) Early steps of lymphocyte activation bypassed by synergy between calcium ionophores and phorbol ester. Nature 313: 318-320
103. Vandesande F, Dierickx K (1976) Immuno-cytochemical demonstration of the inability of the homozygous Brattleboro rat to synthesize vasopressin and vasopressin-associated neurophysin. Cell Tissue Res 165: 307-316
104. Van Loon GR, Scapagnini V, Cohen R, Ganong WF (1971) Effect of intraventricular administration of adrenergic drugs on the adrenal venous 17-hydroxycorticosteroid response to surgical stress in the dog. Neuroendocrinology 8: 257-272
105. Whalley LJ, Borthwick N, Copolov D, Dick H, Christie JE, Fink G (1986) Glucocorticoid receptors and depression. Br Med J 292: 859-861
106. Whalley LJ, Christie JE, Bennie J, Dick H, Blackburn IM, Blackwood D, Sanchez Watts G, Fink G (1985) Selective increase in plasma luteinizing hormone concentrations in drug-free young men with mania. Br Med J 290: 99-102
107. Whalley LJ, Christie JE, Bennie J, Dick H, Sloan-Murphy J, Fink G (1987) Elevated plasma luteinizing hormone concentrations, cryptorchidism and mania. Psychoneuroendocrinology 12: 73-77
108. Whalley LJ, Kutcher S, Blackwood DHR, Bennie J, Dick H, Fink G (1987) Increased plasma LH in manic-depressive illness: evidence of a state-independent abnormality. Br J Psychiatry 150: 682-684
109. Whalley LJ, Rosie R, Dick H, Levy G, Watts AG, Sheward WJ, Christie JE, Fink G (1982) Immediate increases in plasma prolactin and neurophysin but not other hormones after electroconvulsive therapy. Lancet I: 1064-1068
110. Yates CM, Harmar AJ, Sheward J, Simpson J, Rosie R, Fink G, Gordon A (1985) Peptides and amines in Alzheimer-type dementia and Down's syndrome. In: von Hahn HP (ed) Interdisciplinary topics in gerontology, vol 19. Karger, Basel, pp 175-183
111. Yen SSC, Lasley BL, Wang CF, Leblanc H, Siler TM (1975) The operating characteristics of the hypothalamic pituitary system during the menstrual cycle and observations of biological action of somatostatin. Recent Prog Horm Res 31: 321-357
112. Yen SSC, Tsou CC, Naftolin F, Vandenberg G, Ajabor L (1972) Pulsatile patterns of gonadotropin release in subjects with and without ovarian function. J Clin Endocrinol Metab 34: 671-675

Problems and Prospects

George Fink

MRC Brain Metabolism Unit, University of Edinburgh, Department of Pharmacology, Edinburgh, United Kingdom

---

The preceding chapters show that although major advances in our knowledge of the neuroanatomy and neurochemistry of the brain have been made, we still seem to be remote from understanding how the various components of central systems interact to produce co-ordinated sensorimotor responses, thoughts and moods. Thus, perhaps more than any other organ, the brain fits the old adage that the whole is greater than the sum of its parts. The purpose of this chapter is to review briefly possible errors in concepts and techniques and the prospects for future advances in our understanding of brain function and the mind.

Errors and Deficiencies in Concepts

It is trite, but nonetheless true that in spite of the numerous philosophers who have grappled with the nature of mind and reason, we are, after more than 5000 years of recorded history, no further in our understanding of the mind. However, even at a more fundamental and simple level there may be deficiencies in concepts which may impede advances in neuroscience. An example of such a possible conceptual error is our current belief that each particular motor or mental disorder is associated with a defect in one particular neurotransmitter system. Thus, major depression is thought to be associated with dysfunction of the central noradrenergic and serotonergic neurones, while schizophrenia is thought to be associated with major dysfunction of central dopaminergic neurones. Alzheimer – type dementia was thought until recently to be due primarily to dysfunction of central cholinergic neurones. While this type of 'chemical phrenology' may provide useful working hypotheses that can be tested, it does carry the disadvantage that it may distract us from seeing the real source of the dysfunction.

Another example of a misleading concept is the so-called Dale's principle which was accepted and influenced neuroscience thinking and strategy for many years. We now know that neurones are capable of synthesizing more than one transmitter (for details see pp. 7ff.), although it must be admitted that a clear-cut functional role for dual transmitter content and release has so far been shown only for the parasympathetic innervation of the salivary glands [8].

Errors in Techniques and Interpretation of Data

'Chemical phrenology', on which much of modern basic and clinical neuroscience research is based, depends to a large extent on immunohistochemistry (see Chaps. 2, 3, and 4). While immunohistochemistry has proved to be a most powerful tool for investigating central neural pathways and neurotransmission, it has two major potential sources of error. The first is the obvious problem of specificity and cross-reactivity, which most careful workers will do their best to eradicate by well-known and tried methods. However, the problem of shared antigenic sites is more difficult to solve. This problem is exemplified by the fact that an antiserum to the molluscan cardioexcitatory tetrapeptide, Phe-Met-Arg-Phe-NH_2 (FMRF), will also detect the 36 amino-acid residue avian pancreatic polypeptide and neuropeptide Y because all three peptides share a similar antigenic carboxyl terminus [10]. The shared antigenic site caused confusion for some time, in that on the basis of immunohistochemical and radioimmunoassay studies it was thought that all three peptides were present in mammalian brain. Ultimately, only neuropeptide Y could be detected in extracts of mammalian brain [10]. Another example of spurious immunoreactivity is the report that an apparently specific anti-TRH serum detected TRH activity in alfalfa grass, which subsequently proved to be due to the presence of a shared antigenic site in a structural protein of the grass [6].

These types of conformational cross-reaction artefacts can be minimized by cross-checking with antisera raised to several different parts of the molecule and by confirming the presence of the substance by several different chromatographic systems. It is also possible to establish whether a neuropeptide is synthesized in a neurone by determining the presence of the mRNA for the neuropeptide, using in situ hybridization with the aid of a radiolabelled cDNA probe [3, 5]. Neither immunohistochemistry nor hybridization histochemistry are techniques which are easy to use as quantitative methods, but providing the mRNA is sufficiently abundant it is possible to quantify the amount of mRNA present using Northern blots. The cDNA-mRNA hybridization methods are not completely free of problems, in that careful checks must be made to exclude non-specific labelling (as occurs, for example, between certain probes and myelin). Misleading results might also be produced by the presence of fortuitous homologies between the base sequences of the cDNA probe and a mRNA (similar in principle to the problem of amino-acid sequence homologies in the case of immunological methods) [11]. However, in spite of these potential problems, cDNA-mRNA hybridization coupled with careful immunocytochemical, radioimmunoassay and chromatographic methods provides a powerful set of tools to determine whether or not a particular substance is really synthesized in a particular region of the brain. Final proof that the substance is, in fact, a neurotransmitter depends upon satisfying two criteria; i.e. the substance must be shown to be released from nerve terminals and to have an action on its neuroeffector cells which is consistent with its proposed physiological function.

The artefacts produced by so-called receptor studies are as or more frequent and serious than those produced by the various methods used for localization of a neurotransmitter. What is often overlooked, either wittingly or unwittingly, is that a ligand will bind to and be displaced from biologically inert material, and it is therefore crucial to determine that binding has biological relevance and that this binding is 'specific' [1, 14]. Artefacts produced by binding studies can be minimized by determining whether binding actually correlates with a biological response such as a muscle twitch, an action potential, exocrine or endocrine secretion, or an increase in adenylate cyclase activity [1]. The autoradiographic method of detecting receptor localization is potentially important and powerful but must be checked by vigorous mathematical analysis and the use of appropriate positive as well as negative controls [16].

The criticism of the electrophysiological approach are so well known that they have become clichés. Perhaps the best example of the major conceptual criticism is the old story of Martians examining the Houston Astrodome in which a gridiron match is being played. The Martians divide into two teams, the first of which takes sound recordings from the surface of the dome while the second inserts microphones through the surface of the dome. The surface recordings show waves of sound, while the microphones inserted inside the dome pick up sounds such as 'peanuts!', 'popcorn!', 'hooray!'. Neither form of recording would help the Martians to understand what a gridiron match is or the nature of crowd behaviour inside the Astrodome. There are several counterarguments to this conceptual criticism of the electrophysiological approach, the most obvious and cogent being: How else can we ever understand the function of the brain as a whole without understanding its component parts?

Finally, the neuroendocrine-psychoneuroendocrine approach (see pp. 55ff.) illustrates the possible errors inherent in trying to extrapolate from one system to another. Thus, the activity of the tuberoinfundibular neurones in controlling prolactin release may have no relevance to the action of the nigrostriatal dopaminergic neurones. The error of extrapolation is not limited to the neuroendocrine approach, of course, in that assumptions made on the basis of the activity of the nigrostriatal dopaminergic system, for example, may have no bearing on the dopamine systems that project to the cerebral cortex.

Prospects

The prospects for neuroscience research, basic and clinical, are exciting. Notwithstanding the potential errors in concepts and techniques mentioned above, we are now equipped with incredibly powerful instruments with which to investigate the brain. Of the more recent instruments, positron emission tomography (PET) and molecular biology are perhaps the most powerful. PET, by virtue of the fact that it is living autoradiography, which can be used to measure regional oxygen and glucose uptake, blood flow, and the movement of positron-labelled precursors and ligands [12], will probably rewrite neurophysiology, the chemical pathology of neurological and psychological disorders, and the neurochemistry and neuroanatomy of thought and mood. The value of PET has already been demonstrated in certain pathological conditions. Thus, for example, studies of glucose metabolism in Huntington's chorea have identified very early in the disorder a reduction in glucose utilization in the caudate and putamen which precedes bulk tissue loss [7]. The same study identified reduced caudate glucose utilization in a proportion of asymptomatic subjects at risk of developing Huntington's chorea. In young adults with Down's syndrome, a generalized increase in glucose metabolism compared with age-matched control subjects has been reported [15], while in patients with Alzheimer-type dementia, regional reductions in glucose utilization are found which correlate with cognitive deficits [2]. The extension of PET studies to disorders with a less defined organic aetiology has proved particularly rewarding. Reiman et al. [13] have reported a focal abnormality in cerebral blood flow in patients with panic anxiety stimulated by lactate infusion. The focal abnormality, assymetry in blood flow in the parahippocampal gyrus (left less than the right), was not found in patients with panic anxiety who did not respond to lactate infusion or in control subjects.

Molecular biology has already played a major role in delineating the genetic regulation of transmitter synthesis and processing. As a consequence of Numa's elegant studies [9], we now understand to a large extent the structure and function of the acetylcholine receptor, and it is only a matter of a time before many of the questions of neural structure and function will be solved by molecular-biological techniques. Perhaps the most instructive example of the application of molecular biology to clinical neuroscience is the finding of a genetic probe which can detect a specific polymorphic DNA marker that is linked to Huntington's chorea and that maps to human chromosome 4 [4]. The discovery by Gusella et al. [4] was made easier by the fact that Huntington's chorea is an autosomal dominant disorder and that the investigators were able to obtain lymphoblastoid cell lines in the USA and Venezuela from large families with well-established pedigrees for Huntington's chorea [4]. Gusella and associates [4] were also fortunate in finding a probe so rapidly. As well as providing a marker which may prove invaluable for genetic counselling (Huntington's chorea does not manifest itself until the third to fifth decade), the chromosomal localization of the gene for Huntington's chorea is the first step in using recombinant DNA technology to identify the primary genetic defect in this disorder. The principles of the method have been and are being applied to other conditions [17, 19]. Although none of the functional psychoses are inherited as a simple mendelian dominant, there is good evidence that genetic factors play an important role in schizophrenia and affective disorders. With respect to the latter, for example, Wright and associates [18] examined the binding of ^{125}I-iodohydroxybenzylpindolol to β-adrenoreceptors in lymphoblastoid cell lines from members of five families affected by manic-depressive psychosis. Cell lines from six manic depressives, seven unaffected relatives and 11 non-psychiatric controls were examined. Binding was reduced to less than half of control values in cell lines from four of the six manic-depressives and only one of the 18 unaffected relatives or controls. All the cell lines with reduced β-adrenoreceptor binding came from three families; members of the remaining two families showed normal binding. These findings suggest that genetic heterogeneity is present in manic-depressive psychosis and that a β-adrenoreceptor defect may influence genetic susceptibility to the disorder. The studies of Wright et al. [18] thus support the view that there is a genetic element in the aetiology of affective disorders and that, therefore, molecular genetic techniques may prove useful in the diagnosis of such disorders and in the identification of their cause.

Even though the results of psychoneuroendocrine studies (pp. 55ff.) and the prospect of future studies with PET and molecular biology seem exciting, further major advances in our understanding of the biological basis of mind and mental disorders may, in fact, require a step difference in the way we think about the brain, the mind and central neurotransmission. Such step changes in understanding are few and far between. As I mentioned in Chap. 1, 1400 years were to elapse before Galen's hypotheses were overturned by Vesalius, Harvey, Willis and Lower. A factual and conceptual revolution such as that which occurred in the 17th Century may be necessary before we arrive at a rational and perfect understanding of the neurochemical and physical basis for the normal and abnormal functions of the mind. It is at once daunting and exciting to imagine that perhaps in 50, 100 or 1400 years another Richard Lower will overthrow the theories of the present-day chemical phrenologists and synaptologists in the same way as we dismiss the importance of the cerebral ventricles with which the ancients were so obsessed.

References

1. Cuatrecasas P, Hollenberg MD, Chang K-J, Bennett V (1975) Hormone receptor complexes and their modulation of membrane function. Recent Prog Horm Res 31: 37-94
2. De Leon MJ, Ferris SH, George AE, Resiberg B, Christman DR, Kricheff II, Wolf AP (1983) Computed tomography and positron-emission transaxial tomography evaluations of normal aging and Alzheimer's disease. J Cereb Blood Flow Metab 3: 391-394
3. Gee CE, Chen C-LC, Roberts JL, Thompson R, Watson SJ (1983) Identification of pro-opiomelanocortin neurones in rat hypothalamus by in situ cDNA-mRNA hybridization. Nature 306: 374-376
4. Gusella JF, Wexler NS, Conneally PM, Naylor SL, Anderson MA, Tanzi RE, Watkins PC, Ottina K, Wallace MR, Sakaguchi AY, Young AB, Shoulson I, Bonilla E, Martin JB (1983) A polymorphic DNA marker genetically linked to Huntington's disease. Nature 306: 234-238
5. Hudson PJ, Penschow JD, Shire J, Ryan G, Niall HD, Coghlan JP (1981) Hybridization histochemistry: use of recombinant DNA as a homing probe for tissue localization of specific mRNA populations. Endocrinology 108: 353-356
6. Jackson IMD, Bolaffi JL (1983) Phylogenetic distribution of TRH: significance and function. In: Griffiths EC, Bennett GW (eds) Thyrotropin-releasing hormone. Raven, New York, pp 191-202
7. Kuhl DE, Phelps ME, Markham CH, Metter EJ, Riege WH, Winter J (1982) Cerebral metabolism and atrophy in Huntington's disease determined by ^{18}FDG and computed tomographic scan. Ann Neurol 12: 425-434
8. Lundberg JM, Anggard A, Fahrenkrug J, Lundgren G, Holmstedt B (1982) Corelease of VIP and acetylcholine in relation to blood flow and salivary secretion in cat submandibular salivary gland. Acta Physiol Scand 115: 525-528
9. Mishina M, Tobimatsu T, Imoto K, Tanaka K-I, Fujita Y, Fukuda K, Kurasaki M, Takahashi H, Morimoto Y, Hirose T, Inayama S, Takahashi T, Kuno M, Numa S (1985) Location of functional regions of acetylcholine receptor α-subunit by site-directed mutagenesis. Nature 313: 364-369
10. Moore RY, Gustafson EL, Card JP (1984) Identical immunoreactivity of afferents to the rat suprachiasmatic nucleus with antisera against avian pancreatic polypeptide, molluscan cardioexcitatory peptide and neuropeptide Y. Cell Tissue Res 236: 41-46
11. Patient R (1984) DNA hybridization - beware. Nature 308: 15-16
12. Phelps ME, Mazziotta JC, Huang S-C (1982) Study of cerebral function with positron computed tomography. J Cereb Blood Flow Metab 2: 113-162
13. Reiman EM, Raichle ME, Butler FK, Herscovitch P, Robins E (1984) A focal brain abnormality in panic disorder, a severe form of anxiety. Nature 310: 683-685
14. Roth JC, Kahn R, Lesniak MA, Gorden P, De Meyts P, Megyesi K, Neville DM, Gavin JR, Soll AH, Freychet P, Goldfine ID, Bar RS, Archer JA (1975) Receptors for insulin, NSILA-s, and growth hormone: applications to disease states in man. Recent Prog Horm Res 31: 95-139

15. Schwartz M, Duara R, Haxby J, Grady C (1983) Down's syndrome in adults: brain metabolism. Science 221: 781–783
16. Shivers BD, Harlan RE, Morell JI, Pfaff DW (1983) Absence of oestradiol concentration in cell nuclei of LHRH-immunoreactive neurones. Nature 304: 345–347
17. Weatherall DJ (1982) The new genetics and clinical practice. HM Queen Elizabeth The Queen Mother Fellowship. Nuffield Provincial Hospital Trust, London
18. Wright AF, Crichton DN, Loudon JB, Morten JEN, Steel CM (1984) β-Adrenoreceptor binding defects in cell lines from families with manic-depressive disorder. Ann Hum Genet 48: 201–214
19. Editorial (1984) Molecular genetics for the clinician. Lancet i: 257–259

Current Topics in Neuroendocrinology

Editors:
D. Ganten, Heidelberg
D. Pfaff, New York

Editorial Board:
Y. Arai, Tokyo
K. Fuxe, Stockholm
H. Imura, Kyoto
B. Pickering, Bristol
G. Stock, Berlin

Springer-Verlag
Berlin Heidelberg New York
London Paris Tokyo

Volume 7

Morphology of Hypothalamus and Its Connections

1986. 94 figures. VI, 314 pages. ISBN 3-540-16919-9

In complicated tissues such as the parts of the nervous system which govern endocrine responses, morphological studies are more necessary than ever to guide physiological and chemical research. This volume treats the morphology of the hypothalamicpituitary unit not in the way found in traditional textbooks, but in a manner which includes the most modern and detailed structural techniques. Chapters cover not only revolutionary new findings about pituitary and hypothalamic blood supply, but also ultrastructural and histochemical investigations of the anterior pituitary and the median eminence. Connections of hypothalamic cells, ultrastructure of peptideproducing neurons and synaptogenesis in the hypothalamus are also covered. Such topics form the basis on which detail electrical, endocrine, and neurochemical experiments as well as clinical ideas can be planned.

Volume 6

Neurobiology of Oxytocin

1986. 38 figures. X, 175 pages. ISBN 3-540-15341-1

This volume reviews the increasingly important role of oxytocin in the regulation of various biological functions that goes beyond the classical role of this peptide and its effects on lactation and labour. This thoughtfully structured book includes new aspects on the molecular biology and biosynthesis of oxytocin into specific peptide fragments with differential biological activity. Other chapters are devoted to the regulation of oxytocin release from the pituitary; the effect of oxytocin on behavior and memory as well as newly discovered sites of synthesis of oxytocin, such as given in the ovary, and their local effects.

Volume 5

Actions of Progesterone on the Brain

1985. 61 figures. V, 216 pages. ISBN 3-540-13433-6

Contents: *T. Rabe, L. Kiesel, B. Runnebaum:* Antiprogestins. – *J. Kato:* Progesterone Receptors in Brain and Hypophysis. – *Y. Sakuma:* Effects of Estrogen and Progesterone as Revealed by Neurophysiological Methods. – *B. Parsons, D. Pfaff:* Progesterone Receptors in CNS Correlated with Reproductive Behavior. – *P. Söderstein:* Estradiol-Progesterone Interactions in the Reproductive Behavior of Female Rats. – *S. A. Sanders, J. M. Reinisch:* Behavioral Effects on Humans of Progesterone-Related Compounds During Development and in the Adult.

Current Topics in Neuro-endocrinology

Editors:
D. Ganten, Heidelberg
D. Pfaff, New York

Editorial Board:
Y. Arai, Tokyo
K. Fuxe, Stockholm
H. Imura, Kyoto
B. Pickering, Bristol
G. Stock, Berlin

Springer-Verlag
Berlin Heidelberg New York
London Paris Tokyo

Volume 4
Neurobiology of Vasopressin

1985. 53 figures. V, 203 pages. ISBN 3-540-11351-7

Contents: *D. Richter:* Biosynthesis of Vasopressin. – *G. Clarke, L. P. Merrick:* Electrophysiological Studies of the Magnocellular Neurons. – *M. J. McKinley:* Volume Regulation of Antidiuretic Hormone Secretion. – *W. Rascher, R. E. Lang, Th. Unger:* Vasopressin, Cardiovascular Regulation and Hypertension. – *A. Weindl, M. Sofroniew:* Neuroanatomical Pathways Related to Vasopressin.

Volume 3
Central Cardiovascular Control

Basic and Clinical Aspects

1983. 71 figures. V, 192 pages. ISBN 3-540-11350-9

"A group of 18 experienced contributors have combined to provide a thorough up-to-date account of past acquisitions and present research in an important field of physiology, pathology and pharmacology. All important aspects of the subject are covered... Eclectic illustrations and graphs and a large number of references support the text. Teachers and researchers in the broad field of circulation will find this book a very useful source of information as well as a valuable reference."

Archives internationales de Pharmacodynamic et de Therapie

Volume 2
Adrenal Actions of Brain

1982. 25 figures. V, 153 pages. ISBN 3-540-11126-3

"... this volume collects most of the key points and references covering the relationship between the adrenal and the brain with regard to glucocorticoid hormone action... an excellent book which gathers several reviews not previously available in a single volume." *Quarterly Journal of Experimental Physiology*

Volume 1
Sleep

Clinical and Experimental Aspects

1982. 47 figures. VII, 129 pages. ISBN 3-540-11125-5

"The book comprises 4 chapters on important topics in sleep research: both the role of dopamine in REM-sleep and the function of peptides in the sleep-waking cycle are dealt with; circadian rhythms and homeostasis as well as haemodynamic changes during sleep, are similarly described. These 4 chapters are critical, informative... and constitute good reviews." *Acta Neurologica Scandinavia*

If you have any concerns about our products,
you can contact us on
ProductSafety@springernature.com

In case Publisher is established outside the EU,
the EU authorized representative is:
**Springer Nature Customer Service Center GmbH
Europaplatz 3, 69115 Heidelberg, Germany**

Printed by Libri Plureos GmbH
in Hamburg, Germany